'This book will be hard to ignore.'

John Friedmann, *University of British Columbia, Canada*

'Appropriately critical, this wide-ranging and well balanced collection moves beyond the simplistic caricatures of urban regeneration as all good or all bad that have dominated the literature for so long. In so doing it keeps open the possibilities of urban regeneration for creating a socially just city.'

Loretta Lees, *King's College London, UK*

'This is an exciting and thought-provoking collection. It combines a critical review of the international practice of urban renaissance and urban regeneration, with a profound and sympathetic understanding of local experience. It's exciting because it brings together evidence and ideas from across the world and thought-provoking because it points to a range of alternative futures. It is explicitly focused on the ways in which contemporary urban policy helps to generate inequality, but also looks for ways in which dominant approaches can be challenged. The stories told in its case studies are stories of possibility as well as stories of top-down neo-liberalism. Instead of presenting policy as something that is simply handed down to a passive population, these stories offer the prospect of a world in which active engagement can generate positive outcomes. This is a book that should be read by planners and policy-makers, academics and activists, students and teachers. It undermines old certainties and encourages new ways of thinking about old problems.'

Allan Cochrane, *Open University, UK*

'This is an extraordinary and much-needed collection. Porter and Shaw have assembled a truly international cast of critical urban scholars, and their editorial skills have resulted in a book that will surely become the definitive resource for anyone interested not just in the dark side of urban regeneration, but – more importantly – what might be done about it. Any quick skim of the book will be arrested by essays that are truly riveting. The courage, the honesty, and the genuine optimism of the contributors that there can be something other than gentrification will create and enliven debates for years to come.'

Tom Slater, *University of Edinburgh, UK*

Whose Urban Renaissance?

The desire of governments for a 'renaissance' of their cities is a defining feature of contemporary urban policy. From Melbourne and Toronto to Johannesburg and Istanbul, government policies are successfully attracting investment and middle-class populations to their inner areas. Regeneration – or gentrification as it can often become – produces winners and losers. There is a substantial literature on the causes and unequal effects of gentrification, and on the global and local conditions driving processes of dis- and re-investment. But there is little examination of the actual strategies used to achieve urban regeneration – what were their intents, did they 'succeed' (and if not why not) and what were the specific consequences?

Whose Urban Renaissance? asks who benefits from these urban transformations. The book contains beautifully written and accessible stories from researchers and activists in 21 cities across Europe, North and South America, Asia, South Africa, the Middle East and Australia, each exploring a specific case of urban regeneration. Some chapters focus on government or market strategies driving the regeneration process, and look closely at the effects. Others look at the local contingencies that influence the way these strategies work. Still others look at instances of opposition and struggle, and at policy interventions that were used in some places to ameliorate the inequities of gentrification. Working from these stories, the editors develop a comparative analysis of regeneration strategies, with nuanced assessments of local constraints and counteracting policy responses. The concluding chapters provide a critical comparison of existing strategies, and open new directions for more equitable policy approaches in the future.

Whose Urban Renaissance? is targeted at students, academics, planners, policy-makers and activists. The book is unique in its geographical breadth and its constructive policy emphasis, offering a succinct, critical and timely exploration of urban regeneration strategies throughout the world.

Libby Porter is a lecturer in planning in the Department of Urban Studies at the University of Glasgow.

Kate Shaw is a research fellow in the Faculty of Architecture, Building and Planning at the University of Melbourne.

Routledge Studies in Human Geography

This series provides a forum for innovative, vibrant, and critical debate within Human Geography. Titles will reflect the wealth of research which is taking place in this diverse and ever-expanding field.

Contributions will be drawn from the main sub-disciplines and from innovative areas of work which have no particular sub-disciplinary allegiances.

Published:

Whose Urban Renaissance?

An international comparison of urban regeneration strategies

Edited by
Libby Porter
and
Kate Shaw

Routledge
Taylor & Francis Group

LONDON AND NEW YORK

First published 2009
by Routledge
2 Park Square, Milton Park, Abingdon, Oxon OX14 4RN

Simultaneously published in the USA and Canada
by Routledge
711 Third Avenue, New York, NY 10017

*Routledge is an imprint of the Taylor & Francis Group,
an informa business*

Typeset in Times New Roman by
Book Now Ltd, London

First issued in paperback 2013

British Library Cataloguing in Publication Data
A catalogue record for this book is available from the British Library

Library of Congress Cataloging in Publication Data
Whose urban renaissance?: an international comparison of urban
regeneration strategies/edited by Libby Porter and Kate Shaw.
        p. cm.—(Routledge studies in human geography; 26)
"Simultaneously published in the USA and Canada by Routledge."
Includes bibliographical references and index.
Urban renewal—Case studies. 2. City planning—Case studies. 3. Cities
and towns—Case studies. I. Porter, Libby II. Shaw, Kate
HT170.H53 2008
307.3'416—dc22   2008028332

ISBN13: 978–0–415–86071–0 (pbk)
ISBN13: 978–0–415–45682–1 (hbk)
ISBN13: 978–0–203–88453–1 (ebk)

Contents

Illustrations

Figures

Tables

Contributors

Nasser Abu-Anzeh graduated from the College of Architecture at the University of Florence, Italy, and is currently a faculty member at the College of Architecture and Planning at King Saud University, Saudi Arabia. He has also taught at Florence University and Al-Balqa Applied University, Jordan. Mr Abu-Anzeh's research interests focus on environmental and urban sustainable development, affordable housing and public participation. He has been involved in many consultancies for the public and the private sector, and leads the Jordanian research group within the joint Italo-Jordanian Interdisciplinary Research Team for Planning and Anthropological Studies in the South East Mediterranean Center.

Giovanni Allegretti is an architect, planner and senior researcher at the Centre for Social Studies of the Coimbra University, Portugal. Here he coordinates the PhD program 'Democracy in the twenty-first century', and the Observatory on participatory practices. He studied in Brazil, Denmark and Japan. Expert of the European Commission (URBACT Programme), he teaches town management at Florence University, in which he deals with participatory budgets and interactive planning. On this subject he has written several books and organized or taken part in training courses in Italy, Portugal, United Kingdom, Greece, Sweden, France, Brazil, South Africa, Canada, Indonesia and Spain.

Ingo Bader is a geographer from Berlin and writing his dissertation as DFG-fellow of the Transatlantic Research Program Berlin/New York, Center for Metropolitan Studies/TU Berlin on creative cities as a new urban regime and urban cultural economy in Berlin and New York. His research interests are urban economy, urban restructuring, regime theory and culture. He is co-editor of the book *Der Sound der Stadt. Musikindustrie und Subkultur in Berlin* (The sound of the city. Music industry and subculture in Berlin).

Matthias Bernt is a political scientist specializing in urban governance and local politics and working at the Helmholtz Centre for Environmental Research-UFZ in Leipzig, Germany. The main focus of his work is on decision-making in urban politics, with recent projects observing problems of governance in shrinking cities. He has published numerous articles and three monographs on questions of urban renewal, local politics and gentrification. He teaches at the University of Leipzig and is a winner of the Young Talent Award of the German Association for Political Science 2004.

Martin Bialluch is a geographer and musician from Berlin. He works as a free-lance journalist with emphases on economy, development politics and culture.

Daniel Blumer is a geographer and sociologist. He is active in the independent centre *Reitschule* in Berne and is a long-term inhabitant of a self-governed tenant association. He works as a research assistant at the Institute of Social Planning and City Development at the University of Applied Sciences of Northwestern Switzerland. His current research projects focus on the use, importance and perception of green areas in the city of Basel, public space, and urban development along the Subway 5 in Zürich.

Jordi Bonet-Martí is a graduate in philosophy from the University of Barcelona and a PhD student in social psychology at the Universitat Autonoma de Barcelona. His research includes urban social exclusion, local policy and transformations in the labour market. He currently works as a researcher for the Institute of Government and Public Policies on a project about urban social exclusion.

Constance Carr is a graduate of the Faculty of Environmental Studies at York University, Toronto, and is now a PhD candidate in the Department of Urban and Regional Sociology, Institute of Social Sciences, at the Humboldt University, Berlin. Her research interests span urban political economics, social movements, geographies of power and difference and social spatial theory.

Carlo Cellamare is professor of town planning at the Department of Architecture and Town Planning of the University La Sapienza, Rome, and scientific coordinator of several research projects. The main topics of his research are planning and design as complex socio-cultural processes, urban practices, sense-making in town planning, participatory processes and counter-cultural dynamics. He is involved in action research in Rome, and is coordinator of the first Municipality 'City Center'.

Peter Cohen is an urban geographer, land-use planner, teacher, tour guide and avid San Francisco observer who has been involved in a variety of land-use and housing policy work and community planning initiatives. He has a Masters degree from San Francisco State University where he is also an occasional lecturer in the geography and urban studies departments.

Bob Colenutt is a visiting fellow at the University of Northampton, where he is helping to establish an Institute of Urban Affairs and Sustainable Development. He was recently head of Sustainable Development at Northamptonshire County Council and, before that, head of Regeneration at London Borough of Haringey. Active in the property and planning battles over Docklands and the South Bank in London in the 1970s and 1980s, Mr Colenutt was additionally an elected councillor at the London Borough of Lambeth, and worked for the Greater London Council, Joint Docklands Action Group and the Docklands Consultative Committee.

Laura Colini has a PhD from the Department of Urban and Territory Planning at the University of Florence, and is currently an EU Marie Curie fellow at Bauhaus University, Weimar, where she teaches and does research. She works as consultant for local EU and NGO projects, and is interested in public participation, social inclusion, governance and critical media.

Claire Colomb is a lecturer in urban sociology and European spatial planning at the Bartlett School of Planning, University College London. She studied sociology, politics and urban planning in Paris, London and Berlin. Her research interests include: urban policies, urban governance and planning in a European comparative perspective; gentrification, segregation and social mix in European cities; culture and urban regeneration; and the impact of the European Union on spatial planning practices across Europe.

Marcelo Cruz is currently Chair of the Department of Geography at the University of Wisconsin at Green Bay. He received his PhD from the Graduate School of Architecture and Urban Planning from the University of California, Los Angeles. His scholarly and professional work has focused on inner city redevelopment and urban revitalization processes in North and South American contexts.

Jaap Draaisma has been active in Amsterdam's squatter movement since the 1970s and 1980s. He worked for the Borough of Amsterdam North in the 1990s. He is a co-founder and activist in *de Vrije Ruimte*, the association fighting for more free space in Amsterdam by political lobby, lectures, publications and actions (www.vrijeruimte.nl). He is also the co-founder and director of Urban Resort, a foundation trying to fill the gap between the huge amount of empty office spaces and the demand for cheap workspaces for artists and craftsmen (www.urbanresort.nl).

Beatriz García-Peralta has been a full-time researcher at the Institute for Social Studies of the National Autonomous University of Mexico since 1981. She holds a PhD in economics, obtained an MA in urban development, graduated as an architect from the UNAM Architecture Faculty, and has two diplomas in real estate. Her research areas include housing markets, housing policy, and the link between the private business sector and workers' housing provision in Mexico.

Jamie Gough was involved in the squatting movement in London during the 1970s. He worked on employment issues in low-paid industries in the last administration of the Greater London Council before its abolition. Mr Gough has carried out academic and policy-related research in a Marxist framework in industrial and labour geography, urban and regional political economy and policy, and poverty, as well as writing theoretical work on production, the space economy and the spatiality of the state. He currently lectures in town and regional planning at Sheffield University. Mr Gough's latest book, written with Aram Eisenschitz, is *Spaces of Social Exclusion* (Routledge).

İbrahim Gündoğdu is a PhD candidate in the Department of Political Science and Public Administration at Middle East Technical University, Turkey. He is also on the editorial board of the Turkish academic Marxist journal, *Praksis*. His research interests are in the relations of society and space, the rescaling of state, urbanization processes and labour strategies. He has recently worked on the conflictual tendencies among Turkish capitalist classes on downward rescaling of state–society/economy relations, focusing on the setting up of Regional Development Agencies in Turkey.

Tahar Ledraa is an associate professor at the University of Biskra-Algeria. He graduated from the Institute of Town Planning at the University of Nottingham. His field of research interest ranges from migrant adjustments in urban settings, to low-income housing, to planning for marginalized ethnic minorities and urban regeneration strategies. He has been involved in the study and planning of many regeneration projects in Algeria and Saudi Arabia. He is currently a faculty member at the College of Architecture and Urban Planning at King Saud University, Saudi Arabia.

Ute Lehrer is an associate professor at the Faculty of Environmental Studies, York University. Previously, she was an assistant professor at Brock University, Canada, and at SUNY Buffalo, USA. In the 1990s, she worked at the Swiss Federal University of Technology, Zurich (ETH), as well as being the architecture critic for *Tages Anzeiger Zürich*. Lehrer received a PhD from the University of California, Los Angeles and a Lic.Phil from the University of Zürich, Switzerland. She has published widely on architecture, urban design and planning. Her ongoing research interest is in cities and globalization, where she is concerned with image production in and through the built environment, as well as social and economic justice.

Melanie Lombard is a PhD student in the Department of Town and Regional Planning at Sheffield University, where she is researching place-making in

Mexico. On completing an undergraduate degree in government at the London School of Economics, she worked in social housing in London for several years, followed by completing a Masters in planning policy and practice, which included research in Colombia. She is interested in using qualitative approaches to look at socio-spatial issues in the field of low-income housing, specifically in Latin America, and in contributing to interdisciplinary approaches to this area.

Fernando Martí is a community planner, architect and artist, and has served on the board of several community organizations. He was born in Guayaquil, Ecuador, and has lived in San Francisco since 1992. Mr Martí has a joint Masters degree in architecture and city planning from the University of California, Berkeley.

Marc Martí-Costa is a graduate in sociology from the University of Barcelona and a PhD student at the Universitat Autonoma de Barcelona, where he also received his Masters degree. His research includes youth policies, social exclusion and social movements. Mr Martí-Costa has been a guest researcher at the Universiteit van Amsterdam, University of Durham and Universtat IUAV di Venezia, and is currently a member of the team of social movements and participation of the Institute of Government and Public Policies.

James McKay is a writer and cultural activist who cut his teeth in the voluntary arts sector in his home town of Newcastle upon Tyne, going on to pursue a career as a freelance regeneration consultant. Now resident in East London, he also finds time to travel widely in Europe and to develop a burgeoning reputation on London's performance poetry circuit. His words and vocals grace *Follow On*, the acclaimed second album from art-rock/poetry outfit The Morris Quinlan Experience (Round + Round Records, 2007).

Sebastian Müller studied sociology, history of art and general history, and graduated with a PhD in 1969. He taught as professor of planning theory at the University of Dortmund, in the Faculty of Spatial Planning. He edited two books on the Emscher Park International Building Exhibition, and published books and articles on various subjects of history, including German modern architecture, urban planning, urban regeneration, planning politics and planning theory. He retired in 2005, and is now studying the societal risks of globalizing housing markets.

Núria Pascual-Molinas is a PhD student in economic development and international economics at the University of Barcelona, where she received her Masters degree. She works as an associate professor in the Department of Economics at the University of Girona, Catalonia. Her fields of interest are the political economy of cities, city governance, strategic planning for urban development, labour markets and radical economic theory.

Anna Lisa Pecoriello is a town planner with a PhD in urban, territorial and environmental design. She has been a member of the research group LAPEI (Laboratory of Ecological Design of Settlements of Florence University) since its foundation, and is currently a researcher and professor in the Faculty of Architecture of Florence. Her main interests are the participatory approach in town planning and the relationship between children and the city.

Libby Porter started out as a statutory planning officer in an outer Melbourne municipality, and then moved into a variety of research and policy roles in both local government and state government in Victoria, Australia. She escaped the public service to undertake a PhD at the University of Melbourne, and then moved to the United Kingdom, where she has held academic posts in planning at the Universities of Birmingham and Sheffield. She is currently a lecturer in spatial planning at the University of Glasgow, and maintains interests in planning in postcolonial contexts, urban regeneration, and sustainable development.

Ruth Pringle has been a practising artist for over a decade in the United Kingdom, USA, Belgium and currently France. She graduated from Southern Illinois University M.F.A programme in 2004, specializing in installation, conceptual art and art history. Of constant critical enquiry within both her artwork and writing is the uneasy relationship between contemporary art and contemporary society. Recently, she has begun to work within think-tank situations, which use creative thinking skills towards developing unusual solutions to urban problems. This is an extension of her artistic practice, which she has long described as 'creating extraordinary scenes from ordinary objects'.

Ramon Ribera-Fumaz is a lecturer in the Department of Economics and Management at the Universitat Oberta de Catalunya, Barcelona. He received his PhD from the School of Environment and Development from the University of Manchester. His research gravitates around the urbanization of neoliberalism and its alternatives, the articulation between space, the economy and culture, and state restructuring in Europe.

Mark Saunders is an award-winning independent documentary film maker, media activist and writer (www.spectacle.co.uk). His films, including *Battle of Trafalgar* (on the 1990 poll tax revolt), *The Truth Lies in Rostock '93* (on the rise of fascism in East Germany), *Exodus Movement of Jah People* and *Exodus from Babylon* (about Exodus, the utopian urban collective) have been broadcast in the United Kingdom and internationally. Currently *Quand les Papiers Arrivent...* (2007), a film by and about refugees, is receiving national distribution via all the major trade unions and national broadcast in Belgium. His text on the London Olympics, *The Regeneration Games*, has been published by Mute magazine and Eurozine.

Kate Shaw is a research fellow in the Faculty of Architecture, Building and Planning at the University of Melbourne, Australia, with a PhD in urban planning and a Masters degree in social science. She has a background in alternative cultures and activism. Her research interests include social equity, cultural diversity, housing markets, gentrification, and urban policy and planning. She has received several awards for her publications in these areas and is a regular media commentator and advisor to various Melbourne councils and local campaigns.

Angela Stienen has a PhD in social anthropology from the University of Berne, Switzerland. She conducted research in Medellín, Colombia, on globalization, urban regeneration and peace building, and in Berne on migration, urban development, cultural diversity and inclusion/exclusion. She teaches sociology at the University of Applied Sciences in Berne and is an associate researcher at the Human Rights and Research Institute IPC in Medellin. She is currently the co-director of a research project on religion, state and society for the Swiss National Science Foundation.

Elena Tarsi is an architect and PhD student in urban, regional and environmental design at the Department of Urban and Territory Planning, University of Florence. Her research focuses on the urban politics of informal settlements in developing countries, with specific experience in Brazil. In Salvador she collaborated in 2007 with University of Bahia, UNEB, in a project on participatory planning for the regeneration of an informal area.

Lorenzo Tripodi has a PhD in urban planning, and is a writer and film maker. He leads intertwined careers as a researcher in the urban field and a media artist. Life still has not decided what he will do when he grows up.

Bas van de Geyn is an urban geographer, working for the association *de Vrije Ruimte* in Amsterdam. The association is fighting for more free space in Amsterdam, by political lobby, lectures, publications and actions. In 2007 he worked for Urban Resort, a foundation trying to fill the gap between the huge amount of empty office spaces and the demand for cheap workspaces for artists and craftsmen.

Tanja Winkler lectures at the School of Architecture and Planning, University of the Witwatersrand, Johannesburg, South Africa. She completed a PhD in January 2006 on resident-led and involved regeneration possibilities. She has also worked as a design consultant and a municipal official on the Newtown Cultural Precinct and for the Thames Gateway London Partnership, respectively.

Yi Jing is a Masters student at Wageningen University, the Netherlands. From 2005 to 2006 she organized and took part in the various activities of www.oldbeijing.org, including being one of the first members of the Oldbeijing

photo record ing team. Jing is now in the Netherlands as a contact perso
for www.oldbeijing.org in Europe, in charge of cultural exchange and
communication.

Iacopo Zetti is a town planner with a PhD in urban, territorial and environmen-
tal design. He has been a member of the research group LAPEI (Laboratory of
Ecological Design of Settlements of Florence University) since its foundation.
He has worked as a consultant on various design and research projects in Italy,
Africa (Eritrea, Niger), Cuba and China, and has been a researcher and professor in the Faculty of Architecture of Florence. A freelance researcher, his
main interests are the participatory approach in town planning, mapping and
GIS applied to the study of urban and territorial heritage, and transformation
of urban spaces.

Acknowledgements

We owe thanks to many people who assisted us in preparing this book. Our friends in the International Network for Urban Research and Action (INURA) provided inspiration, networks to many of the authors, space to meet to discuss the book, and warm friendship. The British Academy, the Australian Academy of the Social Sciences in Australia and the Australian Academy of the Humanities funded a symposium in Sheffield in July 2007 where a number of the authors were able to come together to discuss their stories. The symposium had additional funding from the University of Melbourne and in-kind support from the Department of Town and Regional Planning at the University of Sheffield. Our thanks also to Leonie Sandercock, John Friedmann, Margo Huxley, Ruth Fincher and to our respective beloveds. Finally, thanks to Jill, in whose lovely little London garden this book was planned in the summer of 2006.

1   Introduction

Kate Shaw and Libby Porter

The desire of governments for a 'renaissance' of their cities is a defining feature of contemporary urban policy. From Melbourne and Toronto to Johannesburg and Istanbul, government policies are succeeding in attracting investment and middle-class populations to their inner areas. The benefits of reinvesting in city spaces made redundant or derelict by years of neglect and abandonment are clear enough. There are good arguments for 'regeneration' even in areas that, while still vital, have experienced sustained disinvestment. But regeneration has a dark side. All the jobs and activity and improvements to the built and natural environments brought by successful regeneration bring, in turn, increases in land values, which can cause displacement or exclusion of lower income users of that place. The effects of regeneration reinforce themselves: the greater the increases in land value, the greater the potential for social exclusion.

The negative consequences of urban regeneration strategies – intended or unintended – are relatively unexplored by their advocates. This is not surprising: the beneficiaries of major projects are hardly likely to advertise their disadvantages. We hear few reflections from politicians, city boosters and property developers on the immediate and long-term effects of elimination of low-income people from city centres. There is little evidence in government decision-making of recognition that urban regeneration affects different people differently. Policy-makers rarely display understandings of the social, economic, cultural, environmental and political complexities of urban regeneration. Nonetheless, urban regeneration has become conventional wisdom within many governments and 'off-the-shelf' regeneration policies are being rolled out in city after city in an effort to catalyse the revalorization of urban space. Egged on by celebrity academics such as Richard Florida (2002), governments and markets are implementing formulaic urban regeneration strategies as though they have universal application and no qualifying repercussions. As cities all over the world are now seeking their 'renaissance', there is an urgent need to critically assess the nature, impact and meaning of this phenomenon.

Whose Urban Renaissance? examines contemporary urban regeneration strategies in Europe, North and South Americas, the Middle East, Asia, Africa and Australia. We are looking for patterns and differences – does urban regeneration premised on major projects and events always produce building forms and uses of little value to low-income users of city space (Searle and Bounds 1999; Searle 2002; Balibrea 2004; Hall 2005)? Do regeneration strategies designed to increase tourism have similar effects on the locals no matter where in the world they are (Zukin 1995)? Do 'creative city' strategies produce a different kind of regeneration, or are they a complement to 'business as usual' (Peck 2005; Shaw 2006; Atkinson and Easthope 2007)? Is urban regeneration always associated with the selling or marketing of place? Who benefits from the implementation of these strategies?

Before going further, we will clarify the words we are using as their meanings are crucial to the discussion that follows.

Urban regeneration is an elastic term. As Allan Cochrane (2007) observes:

> The definition of the 'urban' being 'regenerated' and, indeed, the understanding of 'regeneration' have varied according to the initiative being pursued, even if this has rarely been acknowledged by those making or implementing the policies. So, for example in some approaches, it is local communities or neighbourhoods that are being regenerated or renewed (learning to become self-reliant). In others, it is the urban economies that are being revitalised or restructured with a view to achieving the economic well-being of residents and in order to make cities competitive. In yet others it is the physical and commercial infrastructure that is being regenerated, in order to make urban land economically productive once again. And there has also been a drive towards place marketing (and even 'branding'), in which it is the image (both self-image and external perception) of cities that has to be transformed
>
> (pp. 3–4)

All these aspects are indeed the subjects of vaguely defined urban regeneration strategies, in various combinations and sometimes all at once, as the chapters in this book show. But there is broad coherency in the term. We use regeneration to mean, simply, reinvestment in a place after a period of disinvestment. We refer to regeneration *strategies* as the various mechanisms by which regeneration occurs, whether state or market driven. Regeneration *policies* are particular public policies (no matter how ill-defined) implemented by governments to achieve this.

Urban regeneration is often seen as a euphemism for *gentrification*. We agree that the two can mean the same thing, but they do not necessarily. We accept the broad definition of gentrification as a 'class remake' of the city (Smith 1996), involving the revalorization and 'production of space for progressively more affluent users' (Hackworth 2002:815) and requiring displacement or exclusion of lower status residents, businesses and other users of that place. This process too can be driven by the state or by the market. There is an argument that regeneration should be seen as carrying the same class character as gentrification (Slater 2006). We don't entirely accept this assessment. Regeneration, as a process of reinvestment, may or may not intend (or bring about) such a transformation. We see regeneration and gentrification as occupying different spaces on a continuum of social and economic geographic change, where maximum disinvestment, or 'filtering', is at one extreme, and 'super-gentrification' – where corporate executives displace university professors (Lees 2003a) – is at the other. Regeneration becomes gentrification when displacement or exclusion occurs. The concept of 'exclusionary displacement' (Marcuse 1985) is important here: if people are excluded from a place they might have lived or worked in or otherwise occupied had the place not been 'regenerated', then we regard this as gentrification as much as had they been directly displaced.

Regeneration can have benevolent overtones, then, and it can also mark the beginning of gentrification. Sometimes there is little distinction to be made between the intents behind regeneration and gentrification strategies, and we allow the authors in this book a certain blurring of the terms to reflect this.

Urban renaissance, in contrast, is an expression with no real content at all, used loosely and uncritically by its usually neoliberal advocates to refer to a desired re-emergence of cities as centres of general social well-being, creativity, vitality and wealth. Including environmental concerns about urban sprawl, and recognition of the benefits of more 'compact cities', urban renaissance encapsulates a confusion of ideals of social, cultural, economic, environmental and political sustainability. We use it here in parody of the all-encompassing range of its common usage.

Reclaiming regeneration: the contribution of this book

This book is situated at the cross-roads of urban geography, urban policy and planning. There is a wealth of literature detailing the global economic and social restructuring that precipitated the withdrawal of investment from cities in the twentieth century (for some of the best examples, see Harvey 1985; Fainstein 1993; Sassen 1994; Smith 1996), so we do not explain these processes here. There is a rich field on the causes and effects of gentrification (for summaries and syntheses of the various perspectives, see Rose 1984; Lees 1994; van Weesep 1994; Shaw 2002; Atkinson 2003; and Lees *et al.* 2008), so neither do we cover this ground.

Our starting point is a curiosity about the prevalence of urban regeneration strategies throughout the world. In order to examine the phenomenon more closely, we asked researchers from 21 cities how these strategies are being used in the places they live. Each of the cities has at some point experienced disinvestment and more recently undergone a process of reinvestment. We asked the researchers to detail the mechanics of the process and discuss its effects. The resulting chapters are short accounts of specific regeneration strategies at work in particular places. Some of these are market led; most are driven by government policies. The chapters look at the intent of the strategy: was it to 'improve' the social demographics of a neighbourhood by replacing low-income with higher income people? Or did it lift the median income by providing decent housing, jobs and services for existing residents? They look at the consequences of the strategy: was there any displacement (of residents, businesses, other users)? Who benefited and who lost from the transformation?

Some of the chapters tell of violent, revanchist appropriations of low-cost space for high-income users. Others tell stories about strategies that did not come to fruition. In these cases, we want to know, why did they not? What are the implications of this? Others chapters are stories of contest. Where there was opposition to the strategy, what form did it take, who was involved, how successful was it in altering the original (stated) course? Still others look at whether the urban regeneration strategy itself had socially equitable intents, and whether these intents were achieved. Or were there multiple intents, even contradictory ones? Or was there a different policy intervention in play at the same time, attempting to counter the effects of gentrification?

Case studies that tell a different story to the exemplary becoming of a 'creative city' (Landry 2000; Florida 2002) are rarely collected and analysed, especially by those responsible for promoting and implementing regeneration strategies. Far more influential on government decision-making is the long-established public policy literature that focuses on the drivers of inner city decline, the need for regeneration (gentrification) of derelict neighbourhoods, and the policy tools available to governments to achieve this (Donnison and Middleton 1987; Bianchini and Parkinson 1993; Carley and Kirk 1998; Roberts and Sykes 2000). This body of work neither evaluates the differing impacts of such policies, nor questions the assumptions about knowledge and place that underpin them. It therefore fails to assess what has been lost in the process of regeneration (Porter and Barber 2006).

Many regeneration policies are based on a logic about cities in which a lack of middle-class presence (residents, investors, visitors) is a 'problem'. These areas, according to this logic, become marked by deprivation and disadvantage (Atkinson 2002; Seo 2002). Deprivation and urban decline are depicted as improper parts of urban life, requiring usually state-led intervention to eradicate them from the city and the city's image (Baeten 2004) – especially if that city is positioning itself in the global marketplace of city competitiveness. This logic begs the question, following Cochrane (2007), of why a particular problem is identified as requiring policy intervention at a particular time, and further, why it is identified as an *urban* problem requiring an intervention into urban land.

We propose that the neoliberal policy dogma of urban renaissance has two defining characteristics: first, it constructs urban places as 'in decline' and in so doing targets for 'renaissance' those neighbourhoods with the most vulnerable populations. Second, it produces policies with the aim of restructuring and revalorizing urban space, where success is measured primarily by the rise in land values. That there is particular unanimity among politicians, policy-makers, city boosters, developers and property owners – that is, those who have most to gain from regeneration – is rarely discussed. As Erik Swyngedouw (2007) argues, an apparent evacuation of politics has made much policy-making a highly consensual business.

There is a powerful strand of critical geography, however, that does explore the negative consequences of urban regeneration strategies. The extensive literature in this field takes urban regeneration, along with renewal, revitalization, rejuvenation and of course renaissance, as depoliticized euphemisms for gentrification. All these terms, as Neil Smith (1996, 2002a) points out, omit any connotation of class restructure and displacement. This literature starts from the premise that the transition from a lower to higher socio-economic status population, involving displacement of the former, is precisely what the advocates of regeneration intend. It analyses the strategies employed and emphasizes the various injustices to individuals and social groups that result (e.g. Deutsche 1996; Smith 1996, 2002a; Atkinson 2002; Atkinson and Bridge 2005; Slater 2006; Lees *et al.* 2008). Part of the power of this work is its clarity that processes of disinvestment and reinvestment are a knowing activity on the part of the 'producers of gentrification' (Smith 1987) – the investors, developers, real estate agents, banks, governments and mainstream media – who act, in effect, as the collective initiative behind gentrification.

The drawback of this approach is that it does not allow that regeneration can occur in any other way. It neither allows for different and competing objectives among the producers of urban regeneration, nor does it consider that various injustices might sometimes be unintended. It precludes the possibility of governments acting beyond the interests of the producers of gentrification. We are not certain that they cannot. The state is not always and already repressive: it can be the site of our 'freedoms and our unfreedoms' (Scott 1998:7). Sandercock argues that the role of the state 'is not a given (not simply "the executive committee of the bourgeoisie") but is dependent on the relative strength of social mobilizations, and their specific context in space and time' (1998:102). What of the impact of struggle, of social mobilizations and sustained opposition to gentrification in gaining greater social equity in government policies? What of governments whose decision-making processes are genuinely influenced by housing workers and other progressive and radical activists? The argument that policy has been depoliticized does not take into account systems of governance that cannot comfortably be described as neoliberal. Nor does it acknowledge the opportunities in the existence of competing policies, which create space for productive dissent.

We are seeking alternatives to the virulent gentrification actively promoted in the name of urban renaissance by the neoliberal governments that currently dominate the world stage. We ask openly, can urban regeneration occur without becoming gentrification? Are there cases of reinvestment in disinvested neighbourhoods that include the provision of secure housing, good work and the kinds of services, activities and places that people on low-incomes want and will use? We knew before we began that occasionally there are more egalitarian government approaches. What happens to these intents when they come to be implemented? What are the implications of these different stories? Can we learn from them?

The literature on urban regeneration and gentrification has until recently focused on the major cities of the West, although the flow of stories from smaller and non-Western cities has quickened in the last decade (Scarpaci 2000; Atkinson and Bridge 2005; Badyina and Golubchikov 2005; also Yeoh 2004, though in the language of cosmopolitanism). The increasingly global occurrence of regeneration and gentrification strengthens the grounds for identifying common characteristics, but also suggests possibilities in the contextual specificities.

It is here that the influence of our urban policy and planning backgrounds reveals itself. Part of this book is dedicated to looking for 'what works' in terms of delivering more equitable outcomes. Cochrane observes that 'looking for "what works" is ultimately unlikely to be very helpful in understanding what actually happens, in interpreting the policies that are developed or assessing the initiatives that are launched' (2007:4). We agree that the rush for 'best practice' in urban regeneration strategies has been superficial, often mercenary, and rarely in search of genuine understanding. We are also mindful of Sudjic's warning that 'when one city has what seems like a smart idea for addressing the ills that face it … then it's only a conference away from infecting all its competitors' (2008:11).

The quest in this book is different. The stories of policy initiatives intended to achieve more equitable regeneration examine how they came about, interpret their ability to deliver, and assess their effects. They form a collection of critical case studies that are essential to the mix of methods required to understand processes of urban change (Sandercock 2003; Lees *et al.* 2008). Cities can learn from practices elsewhere. The chapters in this book display a range not just of specific regeneration strategies, but of intents and outcomes. Their collection is not intended as a 'how to', but as a reminder that '(a) there is no thought without utopia, without an exploration of the possible, of the elsewhere; (b) there is no thought without reference to practice' (Lefebvre 2003:182).

The structure of the book

Whose Urban Renaissance? is divided into five parts. The chapters in the first four parts deal with an urban regeneration strategy of one form or other. Part I reveals some breadth in neoliberal urban renaissance policies and their (sometimes devastating) consequences. The geographical sweep of these stories is wide, encapsulating the global South, the Middle East and a rapidly changing Europe. It opens with an interview with Doreen from Silwood, a housing estate in South London that was subjected to a brutal urban regeneration process resulting in disenfranchisement of a large part of her community. Doreen's story effectively summarizes the six that follow hers, with a direct account of top-down regeneration and its effects. The stories are the same; same but different. All involve identification of an 'urban problem'; all involve a state intent on facilitating market reinvestment in space. All involve the transition of that space to a more affluent population and displacement or exclusion of low-income residents. They differ in their treatment of the users of those places, from violent evictions and ethnic/class cleansing to incremental displacement and more 'polite' forms of exclusion.

All the urban regeneration strategies in Part I were contested, but the state was too repressive or the opposition too fragmented to (yet) bring about any real shift. Parts II, III and IV tell different kinds of stories. Part II documents regeneration strategies that did not unfold as anticipated by their architects, due to local limits that somehow modified or mitigated the tendencies established in Part I. It opens with images and a reflection from Brussels, a city often held to be in a perpetual state of marginal gentrification (van Criekingen and Decroly 2003), where two quite different kinds of businesses sit uneasily side by side. How long they will persist in this transitional state is open. The three stories that follow all come from

Germany, though each is quite different. Uneven economic development means each area is experiencing a relative lack of demand from potential gentrifiers. There are questions about what these stories hold for other cities, and perhaps more productively, what the stories from other cities hold for Leipzig, the Ruhr Valley and Berlin. Part II raises fascinating issues about the inflection of the storyteller. When one of the authors joked that maybe 'emphasizing limits is part of our culture', we reflected on the extent to which inclinations for critique are culturally reproduced. Certainly, standards, expectations and assessments vary dramatically across the cultures of the cities in this book. We return to these questions in Part V.

In Part III, the role of local struggle against urban regeneration policies is examined. It opens with a picture story from Beijing, the site of the 2008 Olympics and the scene of the destruction of the city's historical centre. A group of activists in Beijing have been collecting pieces of demolished buildings, in a gentle, tentative demonstration of resistance. The chapters that follow detail potent struggles – *for* urban regeneration in the case of Green Bay Wisconsin, instigated by people tired of being marginalized as 'drunken Indians' and 'white trash'. Other stories detail how local resistance to large-scale regeneration, when catalysed both by a sense of outrage and of possibility, can win real concessions from the state and the market. All these gains were hard fought for, and all were in response to organized demands.

Part IV contains stories of more subtle forms of regeneration and opens with a complex expression of this. The image depicts an old inner city building in Melbourne with a proposed new use: a restaurant serving Aboriginal food and providing hospitality training to local indigenous people (who surely will not be dining there). The chapters that follow offer a rich variety of policy interventions and possibilities in the interests of greater social equity, all of them the result of demands from confident and well-organized activists. Yet none are unqualified. The story from San Francisco perhaps reveals most clearly the ongoing tension: even as the authors question the idea of a 'sweet spot', they suggest it can be found in the continuing struggle.

Part V puts these narratives into conversation with each other and synthesizes the theoretical and policy lessons they offer. Chapter 23 analyses the case studies for what they tell us about regeneration strategies at work: how they are implemented, who is involved, and their consequences. In essence, we answer the question in each story of 'whose urban renaissance?' Chapter 24 broaches the possibility of a radical approach to reinvestment: one that involves urban regeneration without displacement. It addresses those most difficult of questions: can the benefits of reinvestment be harnessed without excluding vulnerable residents? Are such approaches politically viable in the long term?

We urge our readers to critically analyse the stories and consider the workings of the specific policies as well as how they are affected by, and perpetuate or counter, uneven development. Policy and how it works is important because it has crucial consequences. Urban policy is deeply implicated in this most pernicious and inequitable of city processes known as 'urban renaissance'. It is as important now as it ever was for dissent and debate to point the way towards alternatives.

Part I

On urban renaissance strategies

Top down vs bottom up

Doreen from Silwood, a social housing estate in South London

Mark Saunders

Mark Saunders of Spectacle, a London-based independent and participatory media project, has been documenting the regeneration process on the Silwood Estate since 2001 for a feature documentary. This text and the contribution closing Part 4 are based on video interviews with two community activists: Doreen Dower and Jessica Leech. They live on neighbouring housing estates (the Silwood and the Pepys, respectively), separated by a rail track. The two estates were pushed into one Single Regeneration Budget (SRB) regeneration scheme. On the Silwood Estate, the SRB was an entirely Lewisham council-led bid, whereas the Pepys residents, to the authorities' surprise, put in their own 'bottom up' bid.

The Government Office for London insisted they come together as one scheme, to be known as the Silwood SRB. Lewisham Council used the social development 'outputs' of the Pepys regeneration to cover their almost total lack on the Silwood.

The rationale for forcing these two bids together was that Lewisham Council would view the Pepys social development projects as a pilot for Silwood. A small part of the Silwood Estate belongs to another council, Southwark, whose residents, including Doreen, voted for refurbishment rather than demolition and rebuild. On the Lewisham side of the estate, the demolition and rebuild have been accompanied by the transfer of all the council social housing to two 'registered social landlords': London and Quadrant and Presentation housing associations.

Doreen Dower (over leaf) is the Secretary of Silwood Tenants Association (TA). A long-time resident of the Silwood Estate and nearing retirement, she works as a part-time administrator and cares for her husband.

'I think it was always going to be like it is, because I remember ... we had tenants coming to us saying: "they're going to pull down Silwood". Now if you've lived on Silwood as long as I have, every five years they're going to pull down Silwood, so you took no notice of it.

But it turns out this was just before a local election. And the Lewisham councillors at the time were telling tenants, "Oh. It's going to be a garden estate, cottage estate, it's going to be really nice. Houses with gardens". So that came back to the Tenants Association from the tenants, and in the end I wrote letters to say, can you tell us what's going on? ... So that was the first we heard about it. Back to front.

As a Tenants Association person I went to the first meeting, and that was in the summer of that year. It wasn't until the November of that year the neighbourhood

Doreen Dower.

Photograph by Mark Saunders

manager came to our TA meeting and he said, "as of now you are in this SRB bid", and that was the first we heard about it as a full committee.

And that was on the Monday, we had a meeting that week, and then the weekend following that meeting, they were going to do an independent survey. It was basically do you want demolition or do you want refurbishment? – except it wasn't as straight forward as that. But that's basically what they wanted to know.

This independent surveyor came round. I had a vague idea of what it was about. But there is no way that an ordinary tenant would have known what the hell he was talking about.

We chose to stay and be refurbished as against demolition.

And the rest of the Southwark side of the estate chose demolition and the whole of the Lewisham side chose demolition. You've got four high rises, most of them wanted to move. So that obviously tipped it in that favour.

There were things like: we could demolish the four high rises and refurbish the rest; then, one would be total demolition for the whole estate and one would be total refurbishment for the whole estate.

It wasn't a vote, as such, this independent survey merely told the council what they wanted to know. Let's face it. ... We still didn't approve of the way they'd done it because this independent survey was only a percentage of residents. So it wasn't everybody ... but they don't know what they were saying.

So, then Southwark had another separate one for their blocks. ... and they had various pictures of how it could be. So that a bit more information was coming through for the Southwark residents.

And then they were rather upset by the fact that these blocks didn't want to be demolished. So they kept pushing and in the end we went round ourselves and the Tenants Association did another survey, and it still came up with the same, no we want refurbishment. It took about a year but in the end they decided they really had to listen to us, and they said, yeah okay fine, we do the refurbishment.

[Before the regeneration] we had a tenant's hall, a community flat and a community worker at one time. We used to have our meetings in there and we had office space in there. Then there was the youth centre, Five Steps Nursery and a play area.

We used to run the community hall with a premises management committee, all the user groups were on this committee but Lewisham Council were the ones with the purse. They looked after the hall, you know, if it needed repairing.

All the facilities we had on the estate were going to go in one building. I went to meetings and we'd work with the architects to get a building sorted out … and it would have been built in phase one, but it didn't work like that.

They ended up demolishing the community centre three years ago and we still haven't got anything that we can use at this moment.

Silwood did have some projects out [of] the SRB money apart from the demolition side of things, but it was like you had a year's money and then as soon as that was finished that was it; it was dropped and then it might have been money for another year for something else.

They did a business plan, the original business plan, and we went to visit other places, but nothing came of any of it.

Nothing that happened on Pepys [Estate] touched us. In the beginning there was a certain amount of "why have they got that money, it's ours!" Because it wasn't explained to us that when Pepys put in for their SRB bid and Lewisham put in for the Silwood SRB bid, the government said no you can't have two in one place, you got to join them together.

We just thought we were using part of our money for Pepys as well.

So we didn't really get involved with Pepys at all, looking back we should have at least found out what was happening there, and maybe it could have happened on our estate as well.

The whole thing was [Lewisham] council led. And run by SRB people, which are Lewisham people. The minutes were taken by SRB Lewisham people and if something was missed out, it wasn't always put in as an error the following time.

We'd just get used to one lot of people and then one of them would leave and a little while later someone else would come, and so you'd start to get used to them and so it went on. So, definitely, no continuity with people working on the Silwood SRB.

The problem with housing associations is they just focus on their houses. Although they are meant to be non-profit making, they've got to make a profit to carry on building in other places, so they are a business. It's so different now.

On this estate, they've sort of joined up the two housing associations and one is in charge, for instance, of the cleaning and the other one repairs. They're having to treat Silwood rather differently to their other properties. "Cause we kept saying we are one community, we want to be treated as one community."

We were told that it was going to cost the Tenants Association, for sharing an office with another group called Good Neighbours, for the first year 960 pounds, which is half the rent and it's the smallest room, half the rent!

I just said, "well that's us out. Forget it, because we only get a 1000 pounds a year in funding, and it's not for paying rent like that". The only thing I would really like is some sort of office space, some sort of storage space because everything is in my house and I can't move.

I wasn't involved in developing the new business plan at all. We were told at one of the SRB boardroom meetings that they were doing one, and it took ages. I kept asking about it and we never got anything back. Suddenly, it was "oh but it's ready" ….

And we said but we haven't seen it, where is it? In the end, my copy, I got from Janice from the youth centre. Someone obviously gave her a copy and she got it out to me, but that's how I got my copy.

There's been a business plan already. That's thousands of pounds of SRB money wasted, it's never been used. We don't really know what happened in between those two business plans.

At the moment, the community centre is going to be run by London and Quadrant Housing Association. They said right along they weren't going to.

They're having to work at ways of enabling us to use our own facilities. They've now started these meetings with the various user groups. But all that came from our meeting was, "we will help you find ways to generate the money to pay the rent". I'm sorry, but that's not what I want to hear.

That was mentioned, yeah community trust, we could do that (laughs). That's like another hat, isn't it? It's another difficult job that tenants and residents are supposed to do. I went to a meeting about community trusts … I wasn't that enamoured because I thought, "oh God that's a lot of hard work" … It's okay if there's a lot of other people dying to get on the community trust.

But I suppose in their view it's what we always wanted. A community centre. We kept on about it long enough. But that's what we've got, this really "iconic" (pulls face) building with flats above it. It's ridiculous!

Who in their right mind would choose to live and to part buy, not just to rent, a flat above a community hall? The mind boggles.

We've got various projects around that might want to use the hall, but … they can't afford, and it would be unfair to ask them to afford, in my view anyway, those sort of rents, because it's community.

They gave us one of the old shops in Reculver Road [for a temporary office] which was in such poor condition it's unbelievable. We can't get in there at the moment as there was a fire at the back of the shop and it was damaged, we've got a computer which is being stored possibly by the youth club. We've got a photocopier, I don't know whether it still works but that's in the flat, and we've got two filing cabinets in the flat. So where do we go from here? I don't know.

They would have helped if they had built the community facilities at the beginning when they said they would. We might just have been in the position then to have got in there and got stuff moving before everything was pulled down. There was no centre of any description, so the community's lost.

As far as I'm concerned the new Tenant and Residents Association needs help. Even if it's going round knocking on doors, they can't be expected to do it all themselves, because they're starting from scratch, and it's quite a big estate.

I've been to so many meetings, I'm so tired, my brain's ceased, I just don't know anymore.

[Lewisham Council] are still supporting the youth club but they're nothing to do with the community hall. Because it's not theirs, they haven't built it. They've got shot of the whole estate. And I'm sure that's what they wanted from the beginning.

I'd go to a meeting and realize that a goalpost had been shifted, which meant that things on this estate were different to what they were originally going to be. And this kept happening. Various reasons were given for various things being done differently.

So, I feel let down, completely ... so much so that when I read in the newspaper about so and so estate going down the regeneration road, I go, don't go down that road please, because its terrible. But it probably isn't if it's done properly ...

We are still here, but it's so completely changed.

It's a nicer place in terms of houses with gardens, let's face it most people want to live in a house with a garden, but, because there's nowhere for the kids to play, they're in the streets and we've had a lot of problems, so much so that the people in these blocks have voted to stop off the walk ways [connecting the two parts of the estate].

You need a centre of some description, and that's the only one we've got. Quite how we can manage it, I don't know, but whether we're involved with it or not is neither here nor there. Somehow it's got to be made to work.

You've got to bring the community together, and any sort of regeneration destroys the community. I don't care what anybody says.'

2 Class cleansing in Istanbul's world-city project

İbrahim Gündoğdu and Jamie Gough

> We should find a way to keep poor people from the city of Istanbul.
> (Erdoğan Bayraktar, chairman of the Mass Housing
> Administration of Turkey, 2006)

Istanbul is currently being restructured with the aim of becoming a 'world city'. For both Turkish and international capital and for the Turkish state, Istanbul has the potential to become a major financial and business–service centre serving the whole Middle East and linking Turkey into the European Union and global economy. This project involves the transformation of its labour force, resident population and built environment. The national state has embarked on a Faustian restructuring of the built environment of the whole city region, including commercial and industrial activities and housing. In doing so, it has run up against resistance not just from some sections of capital, but also – our interest here – large parts of the lower income population. Istanbul still has major areas of squatter settlements and derelict buildings on high-value land in the inner city. The state is organizing the eviction of most of the residents and the conversion of this land to offices and to luxury housing and shopping malls serving the growing business elite.

There is, then, a sharp contrast to the gentrification of the inner parts of world cities in the First World (Smith 1996; Hamnett 2003): whereas in the latter gentrification has been accomplished principally – though not entirely – through property markets, in Turkey the state has played the leading role. Paradoxically, this flies in the face of neoliberal ideology, which denies any substantial role to the state other than repression. Moreover, the importance the state gives to current urban restructuring is indicated by the fact that it has been the national rather than local government which has taken the lead. Another striking difference from the First World examples is that in Istanbul there has been mass resistance to the eviction of the poor. This chapter aims to sketch explanations of these specificities.

This programme of restructuring the inner city, and the resistance to it, needs to be understood as the product of processes at different scales from the globe to the neighbourhood. In section two, we set the national scene by considering the conflictual and crisis-ridden history of urban housing in Turkey since the 1950s. Around the turn of the millennium, this evolved towards an accelerated (re)development of cities, reflecting the intensified integration of Turkey into the international economy. In section three, we examine a key part of this change, the reconstruction of Istanbul as a 'world city'. We discuss the consequent conflicts around the built environment, particularly the housing of poorer residents. These dramas have been enacted at neighbourhood levels, but are set within this geographically larger frame. In section four, we outline the resistance to the eviction of the poor and draw some political conclusions.

The evolution of policy on urban housing in Turkey

Since the rapid growth of urban population in Turkey in the 1950s, housing has posed major problems. There were, and still are, only four significant housing tenures in Turkish cities: owner occupation with freehold or on squatted land and private rental from either of these. The Turkish state has never constructed housing for low-cost renting, so 'social housing' means low-cost freeholds. For 30 years, the state allowed new immigrants from rural areas to construct *gecekondu* (squatter settlements) on state-owned vacant land in the cities, built either by the settlers themselves or by land speculators. This was, in effect, one of the few redistributive policies of the state in favour of industrial capital since it meant that a labour force could be built up without these new workers making 'excessive' wage demands. In this period, formal state intervention into urban housing was both through subsidized credit to building firms and direct building by the Turkish Real Estate and Credit Bank and the Social Insurance Fund, which provided housing for sale. In any event, these institutions provided little lower-income housing and a considerable amount of upper-class developments, leading to criticism of them even by middle-income people (Buğra 1998).

In the mid 1980s, the government turned to a neoliberal strategy. This resulted in a sharp decline in wages and consequent worsening of the crisis in urban housing. To deal with this, the government introduced amnesty laws, including 'improvement plans', with which existing illegal buildings were not only regularized, as in previous measures, but were given further construction rights. This was accompanied by the devolution of urban planning to municipal authorities, with the aim of stimulating market-driven construction and land development. In this way, many squatter settlements were transformed into authorized low-quality apartment buildings through the agency of small- and middle-scale builders. Ironically, amnesty laws and regularization plans were also used to construct large residential complexes for the upper-class on empty peripheral land, including areas of forests and reservoirs.

A second reaction to the crisis was to reform the state's funding of housing. The Mass Housing Administration (MHA) was founded in 1984 as the central state institution entrusted with encouraging and undertaking the construction of housing projects backed by large-scale state funding. The MHA was more successful than the previous institutions, and in the late 1980s and early 1990s, constructed a large number of housing cooperatives for middle- and low-income people in the peripheral areas of large cities, comprising in all more than 200,000 residential units.

Following the previous practice, though, in this period the MHA also constructed nearly 40,000 units on its own land in Ankara, Istanbul and Izmir for sale to middle- and upper-class people, forming prestigious residential areas. This major role of the MHA, however, was sharply curtailed in 1993 as a result of a severe fiscal crisis, with its main funding stream being switched to non-housing purposes.

In the 1980s and 1990s, aside from these results of state policy, there was further expansion on the city peripheries. This was partly due to new squatter settlements as rural–urban migration continued. It was also the result of upper- and middle-class housing in privately planned projects. In all, there was a strong movement of higher-income people towards the city peripheries, and emergence of a fast-changing pattern of spaces fragmented along class and ethnic lines.

From the beginning of the 1990s to 2003, there appeared 'regeneration' projects in inner city neighbourhoods with existing low-rent uses, but potentially high rent ones. In the capital city Ankara, such projects were directed at certain squatter areas (Özdemir 1996), whereas in Istanbul they targeted particular parts of the historic centre with lower-income residents (Islam 2005). These initiatives have led to gentrification, but the pace has been slow: both socio-economic and political-legal difficulties mean that the projects have been limited in number and small scale.

A series of events around the turn of the millennium led to an investment boom in the urban built environment, comprising commercial property as well as housing, and a much stronger strategy of the national state towards urban restructuring. The first event was the massive earthquake of 1999 in the Marmara region, where leading economic cities including Istanbul are located, producing more than one-third of the country's total output. This was used to attack existing urban policies and planning institutions, hypocritically ignoring their trajectories based on neoliberalism (Şengül 1999), and to legitimize a new discourse of urban regeneration. In 2000, the Turkish state started discussions about entering the European Union. In 2001, the Turkish economy experienced a sharp crisis and recession. The dominant sections of capital responded by opening up the economy further to international capital, especially financial capital (Ercan and Oguz 2007). In 2002, the Justice and Development Party (JDP) came to power for the first time. The JDP, to a considerable extent, represents provincial small and medium capital in manufacturing, construction and commerce. It has a strong orientation towards the European Union and so-called globalization processes within a moderate project of political Islam. This has put it into conflict with the state bureaucracy, the military and the sections of capital linked to them, which together had dominated Turkish politics since the nationalist revolution of Ataturk. The JDP has seen urban restructuring as an essential part of integrating Turkey more strongly into the European Union and global economy – and also, conveniently, as a way of boosting the domestic construction sector.

The main institutional means chosen by the JDP for this aim has not been the municipalities, but the centrally funded and controlled MHA, which has been reinvigorated and expanded greatly in its powers and scope. Although most big city administrations are in the hands of the JDP, the government has regarded them as too weak administratively, technically and financially to undertake the scale of restructuring envisaged. The remit of the MHA has been both to provide better housing opportunities for middle- and low-income people and to initiate projects of luxury housing and associated up-market consumer services. Regarding the former, in its initial 'Emergency Action Plan', the JDP government declared the 'regeneration' of squatter areas and the provision of social housing for low-income groups as its major urban aims (AEP 2003).

The power of the MHA was expanded in four major directions. First, it was given powers to establish companies related to the housing sector and to go into partnership with existing companies, grant credits for, or directly undertake, the transformation of squatter areas and preserve and restore historical and regional architecture. Second, the MHA was empowered to undertake, directly or indirectly, profit-oriented projects in association with the private sector, known as 'revenue sharing', to give an income stream to fund its other activities. Third, the MHA has been given, for the first time, urban planning powers: to make plans at all scales, and alter existing plans, in areas designated for mass housing development and to carry out compulsory purchase of property and land within these areas. Fourth, the government transferred all the duties and powers of the national Urban Land Office, together with its land bank of 64.5 million square metres, to the MHA in order to integrate housing production with land acquisition and development. The MHA has thus become an institution with exceptionally strong powers for the (re)development of urban space, being able to bypass conventional regulations, institutional bodies and plans and create local project agencies operating like private companies.

With this framework, the MHA has initiated a countrywide housing programme, constructing 286,000 housing units during the last five years through contracts with private builders. Of these, 144,000 units have been provided for middle- and low-income people and 61,000 units are supplied to the poor people under long repayment maturities of 20–25 years. Another 40,000 have also been built in areas hit by earthquakes and other disasters. At the same time, the MHA has undertaken revenue-sharing projects involving the construction of 44,000 housing units in luxury residential areas. In the latter, private developers bid to construct housing on state-owned land, and a site-specific contract is agreed under which the MHA allocates the land to the developer and is paid in instalments over 4–5 years a proportion of the actual profit made from selling the houses; the MHA typically appropriates 25–30 per cent of total revenue. The resulting income to the MHA is used to subsidize the middle- to low-income housing, which is sold at below-market prices. The head of the MHA argues that these revenue-sharing projects are a new model of funding for a public authority with a limited budget (Bayraktar 2006:233). In the last five years, the MHA has sold land in large cities in return for $US 4 billion, which amounts to over one-third of its total investment expenditures over the period.

Revenue-sharing schemes are of great benefit for the private developers. Because of the MHA's planning and land assembly powers, urban land is easily acquired and the legal procedures are completed in a short time. The MHA guarantees for buyers the quality of construction of the housing, an important consideration given the generally low standards of the Turkish building industry;

consequently, the houses are strongly marketed even at their project stage. Moreover, the developers carry very little risk since most of their payment for the land occurs after they have sold the houses – as the association of property companies admits (GYODER 2007).

In this context, the inner areas of large cities, including squatter settlements, have increasingly been the strategic focus of the MHA's profit-oriented projects because of their 'rent gap'. We have seen that such projects were undertaken since the early 1990s, but without the state's help they progressed at a slow pace. The MHA has now changed this situation radically, and is responsible for most of the current inner city developments. It has developed a protocol for the 'regeneration' of squatter settlements with the local authorities. The local authorities identify the area for redevelopment, the protocol is signed, and the area is legally defined as an 'urban regeneration zone'. The MHA and local authorities then map together the current property relations. Existing *owners* are invited to either sell to the MHA, or participate in the regeneration process with their own capital. *Tenants*, on the other hand, have no right to stay in their accommodation, but are given a chance to buy a house from the MHA's new social housing projects on vacant areas at the fringes of the cities. Although the latter are subsidized by the MHA, they are still beyond the reach of most of the tenants. In this way, the inner city squatter areas are cleared and redeveloped for luxury housing, with the MHA as well as developers reaping the profit. By the end of 2007, 28,000 units of luxury inner-area housing have been started through this programme, most in the three biggest cities; more than a hundred municipal authorities have applied to the MHA to develop 113,000 further units.

The past 20 years, then, have seen an acceleration of 'the urbanization of capital' (Harvey 1989a) in Turkey, with increasing investment in the built environment and property markets freed from traditional social constraints (Şengül 2003). This has involved major changes in the dominant actors. From the 1950s to the 1980s, small to medium builders predominated, whereas since then it has been national and international builders, developers and finance. In the earlier period, the poor could find housing through using public land, under the clientalist patronage of the district municipalities; now this land is being fully commodified, and control has been shifted up to the greater municipalities and nation-state where the poor have even less influence (Kurtuluş 2007).

Istanbul under the MHA's wing

From the Ottoman period, Istanbul has been the major city of Turkey with regard to economic activities and social dynamics. It was the centre of emergent industrial capital throughout the nationalist, developmentalist era in the aftermath of the Second World War, and experienced massive inward migration. It rapidly spread outwards, particularly through the expansion of squatter settlements near the new factories on the (moving) edges of the city. However, with the adoption of neoliberal strategies from the mid 1980s, the city has entered into a new phase, in which the shock of Turkey's integration into supranational capitalist dynamics is focused on Istanbul. Peripheral squatter settlements have continued to grow, but at the same time emergent globalizing commercial spaces and upper-class residential areas emerge towards the outskirts of the city. Hence, in the greatest metropolis of the country, disparities of income, wealth and power deepen, expressed in spatial segregations (Keyder 2005).

In the last decade, the city has taken a further step in its internationalization. Istanbul has been conceived by Turkish capital and state as a supranational regional economic centre, serving to accelerate the integration of Turkey into global capitalism, superseding Beirut as the financial hub of the Middle East, and linking the latter to Europe. To this end, the JDP government set up a new planning authority for the city, Istanbul Metropolitan Planning (IMP), which operates at a new spatial scale for the state, namely the whole city region. The IMP has strategic planning powers which override the previous, smaller municipalities. It has adopted two essential aims: the decentralization of manufacturing industries towards outer edges of the built-up area and the transformation of the inner city towards finance and business services and up-market consumption and residential spaces, thus moving the growth in the latter uses from periphery to centre. It has proposed large-scale urban (re)development projects as the main tools for this spatial restructuring. These include three large sea ports at Haydarpasa, Galata and Zeytinburnu, which incorporate trade centres, offices and hotels, and which use existing public land, buildings and green spaces. In addition, new sub-centres are to be created in the outer east and west sides of the city to accommodate local, lower-level commercial activities, enabling the inner city to be freed for higher level business sectors.

Redevelopment of the inner city towards the new internationalized uses is taking place partly on vacant land owned by public authorities. It has also been targeted towards rundown residential areas with poor inhabitants; the IMP has seen a 'rent gap' in these areas, and they appear as its major planned 'regeneration' projects. These areas are composed, firstly, of historic buildings formerly occupied by ethnic minorities of all classes who were expelled from the country by the nationalist regime in the 1920s and 1930s, and, secondly, of squatter settlements built by immigrants in the past 50 years. The workers in these areas are mainly employed in informal service sectors in the city centre. About half of the units are owned by the occupiers, while the other half are rented. The ownership structures have, until recently, made it difficult for private developers to enter into these potentially valuable areas of the inner city. In this context, there has appeared a particular division of labour between capital and the state and also national and local levels of state. The MHA has taken a central role: it has provided technical help to the municipalities; on publicly owned land, it has chosen the private developers; and above all, it has, directly or indirectly, carried out the eviction of the existing residents.

A striking case is the recent regeneration project in the Sulukule district, two dilapidated neighbourhoods near to the historic centre which have been occupied by Anatolian Gypsies since Byzantine times. The MHA-led project is aiming to replace existing buildings with 'Ottoman-style' villas; it is projected that more than a hundred buildings are to be demolished. The MHA has offered housing owners two options: either to buy newly built luxury apartments at a high price or to buy housing units provided by the MHA in the peripheries of the city at below-market rates;

the residents who are renting are also given the chance to buy such 'social housing' units. Since the residents are mostly employed in temporary jobs with low and irregular income, these purchases are not easy, especially for those presently renting and thus lacking any capital. Moreover, the MHA's housing area is 45 km away, making access to employment in the city centre extremely difficult and costly.

Another project, Kucukcekmece, is to create large residential developments near a planned new commercial growth pole in the outer west line of the city. The MHA first demolished nearly 2,000 squatters' housing units in the area and transferred their owners to social housing in less valuable areas. The very poor renters in the squatter settlements were again disregarded when they could not afford to buy these units. The MHA has now started construction of 100,000 housing units, an enormously increased density, with consumption and recreational spaces.

Other similar regeneration projects are currently in planning. The latest statements by the MHA imply that these are just a start. 'We will enter into many squatter areas of Istanbul this year', said the chairman Erdoğan Bayraktar, in a recent TV speech. According to the MHA, half of all the housing units in Istanbul, nearly 1.5 million, violate either the development plans or statutory procedures for the building process. This is presented as conflicting with the project of making it a global city, and also as dangerous in the event of another major earthquake. Accordingly, the MHA envisages demolishing approximately 60 per cent of the settled area of Istanbul. It is expected that more than 2.5 million people will have to move to the periphery. The motivation of structural weakness of building appears hypocritical when one notes that squatter settlements located on the city's hills, which are low risk because the ground is less vulnerable to tremors, are a priority for demolition because of their wonderful panorama of the Bosporus. They are targeted for luxury residential areas that would bring the MHA large profits.

Despite the projects now in hand, the state and capital have not been satisfied with the pace of the transformation process. The laws on urban development, even including the recent enlargements of state powers described above, are regarded as being inadequate to provide local and national bodies with decisive authority. Thus, a new law concerning the process of urban transformation has, in February 2008, been prepared by the government with the support of the property capital and municipalities of large cities. Under this law, urban redevelopment projects would not have to conform to city plans, and public authorities would have yet-stronger powers to expropriate buildings (Ekonomist 2007). The MHA chairman declared that a comprehensive attack on squatter areas of Istanbul would start with the introduction of this law. His words tell of the increasing authoritarianism of urban policy in Istanbul: 'The MHA is dependent on the state, not the people, and they must obey the rules. If they don't, the land will be expropriated and developed along profitable lines' (CNNTURK, 14 January 2008).

Resistance and the future

In recent years, however, the MHA's projects have been resisted by residents. As the projects have been prepared through the cooperation of the MHA, municipalities and property companies without any involvement at all of the residents, resistance has generally started with collective attempts to understand what is being proposed, which in a short time are transformed into neighbourhood organizations. Their basic demands have been to halt the demolitions and evictions, and to allow residents' participation in both preparation and implementation of the projects. Forms of resistance have differed between neighbourhoods depending on their social structures, political traditions and (hence) organizational capacities. For example, the districts of Derbent/Sarıyer and Aydost/Pendik have organized as an urban movement for the right to housing in their existing living places. Residents in Gulsuyu and Gulensu districts cooperated with activists, including academics, students and the 'Planning Workshop in Solidarity', to produce an alternative plan for regeneration without evictions. In Sulukule, the Gypsy residents' organizations have denounced the evictions as ethnic cleansing of the inner city, and have brought the issue to the European Parliament as a violation of their ethnic cultural rights.

The neighbourhood movements have also sought the support of political parties, left groups and professional associations. Workers in the inner city squatter areas have been increasingly subject to informal, insecure and casual employment, so direct links to trade unions have been weak, but left sections of the Turkish union movement have recently become interested in urban issues, particularly in Istanbul.

The state, however, has responded with strategies, ranging from subtle to brutal, to weaken the resistance. The authorities have used their access to the media to present the dwellers in squatter areas as 'invaders' of public land, notwithstanding the previous decades of their legitimation. Demolition has been spread over time to divide different groups of residents. Divisions between owners and tenants have been encouraged, for instance, by offering owners preferential access to new housing, and in some cases this has weakened neighbourhood unity of purpose. Public services such as transport, water and electricity have been cut off or closed. Finally, the residents have been frightened by pressures applied to leading resisters, and by police harassment targeted on the most marginalized groups including Kurdish immigrants, left activists, homosexuals and people of African descent.

These attacks have enabled the state, so far, to more-or-less continue with its plans. Resistance has elicited some relatively minor changes to the development process, but has in no case halted or radically changed it. The authorities in some cases have made short delays in demolishing houses, have agreed to complete new social housing for owners before demolition took place, or have given temporary income or rental support to tenants. A necessary condition for obtaining such concessions seems to be tenacious organization and effective use of the media. Unfortunately, Sulukule residents have just failed to prevent demolition and eviction, despite getting some positive coverage in the media and eliciting a certain pressure from the European Union on the municipal authorities. A similar failure has also been experienced in Maltepe-Basibuyuk and Pendik-Aydost even though they established strong neighbourhood organizations.

The movement has come to realize that it has been weakened by its fragmentation between the various neighbourhoods, so that resistances have been uncoordinated in space and in time. As a result, more than 20 neighbourhood organizations have established a new association at city level, the Platform of Istanbul's Neighbourhood Associations (PINA), so as to improve communication and develop solidarity between neighbourhood movements, and to strengthen support from other social actors. Its founding declaration expresses the historical irony of the present 'regeneration' initiatives and their class nature:

> Our story dates back to the 1950s. As we had not been able to live on in our villages and towns due to the lack of investment, we moved to large cities … State and capital encouraged us to be workers in their growing factories, without any social policy on low income housing, [so that] we had to occupy public land …. In spite of living in squatter areas, we created competitive industries and spectacular cities. But as these developed and became involved in spatially-wider networks, we began to be seen as rough workers unworthy to be living in inner cities. The state and companies are now seeking to evict us from our living places.
>
> (PINA 2007:103–4)

Although it lacks experience, PINA provides an important step in coordinating resistance at wider scales. In our view, the future of the movement depends crucially on the creation of alliances with other progressive social actors, especially the trade unions, to develop an active counter-strategy for the whole city region, and thus to draw support from the majority of Istanbul's residents.

3 Believing in market forces in Johannesburg

Tanja Winkler

Renaissance-style policies are 'not a sideshow in the city, but a major component of the urban imaginary' (Ley 2003:2527). Accordingly, 25 years of capital and white flight from downtown Johannesburg recently prompted the local state to implement a plethora of investor-friendly policies to re-attract private capital and middle-class households. Discursive regeneration policies, which deploy carefully selected discourses such as 'economic competitiveness', 'responsive governance' and 'social cohesion' to obviate criticisms of gentrification, are thus not restricted to command centre cities of the global North. They now appear in the global South, imported as 'best practice', 'world class' enabling precedents to facilitate a global age of regeneration (Bourdieu and Wacquant 2001; Gordon and Buck 2005; Smith 2002a).

However, and contrary to global North experiences, decades of capital and white flight resulted neither in a depopulation of Johannesburg's inner city nor in a vacant, boarded-up landscape. Rather, informal socio-economic activities coupled with a significant inward migration of job seekers have transformed Johannesburg's downtown. Today, the majority of existing inner city residents are poor, many rely on the informal sector to survive, and many reside in physically dilapidated apartment blocks, or 'bad buildings' as classified by the city council, while being exploited by slumlords. Informal socio-economic activities and a doubling of the inner city resident population, in particular, are perceived by municipal officials, policy-makers and politicians as undesirable and unmanageable obstacles in achieving their 'world class African city' imaginary (CoJ 2006a, 2007). Consequently, 'renaissance' style policies are seemingly designed to shift undesirable and unmanageable 'obstacles' via eviction and other mechanisms to 'peripheral locations where they are less of an eyesore and [less of] a threat to the City's renewal process' (Silimela 2003:152). This suggests that inner city regeneration in Johannesburg is nothing more than a euphemism for underlying gentrification.

By means of a critical discourse analysis to place 'the gentrification debate into a policy perspective' (van Weesep 1994:74), I will investigate the apparent assumptions underpinning the City of Johannesburg's regeneration policies to generate a fuller understanding of who stands to benefit, and who does not, economically, spatially and politically, from public sector-led 'renaissance' strategies. My understanding of existing residents' economic, spatial and political needs is based on a three-year, in-depth study with local civil society organizations, residents, community development facilitators, and informal traders. This chapter will, therefore, be structured in two sections. The first will analyse the evolution of the municipality's regeneration discourses since 2000, when 'inner city renewal' was first conceptualized as an economic growth strategy. The second section will then evaluate the city's social policies aimed at complementing 'economic competitiveness' and 'responsive governance' agendas. Here, it will become apparent that despite the implementation of social policies, patterns of access to economic, spatial and political resources continue to be unequally distributed, and that the city's blind faith in international 'renaissance' style policies does not suit the Johannesburg context. If such regeneration policies continue to be implemented, then, arguably, the desired 'world class African city' may lead to a gentrified space of social, economic and political exclusion.

Setting the context: the city's 'evolving' regeneration policies

Following the first local democratic government elections held in December of 2000, 'inner city renewal' was declared one of six mayoral priorities by the newly appointed executive mayor of Johannesburg, Amos Masondo. At the same time, policy-makers formulated an economic development framework with a 30-year horizon, known as the 'Jo'burg 2030 Vision'. By 2030, the framework proposes, Johannesburg will be a world class city: 'Its economy and labour force will specialise in the service sector so that the economy will operate on a competitive global scale [to] drive up tax revenues, private sector profits and disposable incomes' (CoJ 2002:3). The 2030 Vision thus laid the foundation for the first 'Inner City Regeneration Strategy' (CoJ 2003a) with an overarching goal 'to raise and sustain private investment leading to a steady rise in property values' (CoJ 2003a:2). Achieving this goal required an overt demonstration by the municipality to accommodate investor needs, while simultaneously embracing 'responsive governance' discourses that shifted the local authority's role from acting merely as 'an administrator to becoming an active agent of economic development and growth' (CoJ 2002:3).

Accommodating investor needs and responsive governance discourses included establishing public/private partnerships, declaring an 18 km² district in the inner city as an Urban Development Zone so that budding investors could become eligible for substantial tax breaks, stimulating buoyant economic development by supporting 'big business' through the design and implementation of carefully crafted physical interventions, investing in catalytic projects that presuppose a multiplier effect of increased property values through complementary private sector investments, and facilitating the Better Buildings Programme (BBP) by writing off arrears on identified 'bad buildings' and transferring the ownership of these buildings to private sector developers for rehabilitation.

R19 billion (US$ 2,290 million) of public sector money will be allocated to ongoing catalytic, or flagship, projects (*Financial Mail*, 22 July 2005, p. 6) so that 'Johannesburg can be marketed as an exciting Afropolitan city' (Gevisser 2004:517). Further, City Improvement Districts are being set up to facilitate the management of these projects. A 'responsive governance' agenda also involves implementing 'intensive urban management' policies comprising, among other initiatives, tenant evictions from buildings earmarked for the BBP.

> In 2003 we issued 309 eviction notices. That's important! The moment a building is empty we secure it for the BBP. Things are beginning to fall into place because we started where we should have started many years ago, with intensive urban management.
>
> (Interview, inner city director, 2004)

Securing 309 evictions in one financial year is of benefit to officials whose annual evaluations and performance bonuses are based on quantifiable key performance targets. 'Often innocent people's rights get trampled during eviction raids', a police spokesperson informs *Financial Mail* readers, but, for the municipality, 'there is, [apparently], no other way to save the inner city from sliding irrevocably into the abyss' (*Financial Mail*, 10 October 2003, p. 13). Intensive urban management also results in 'building after building facing water and electricity cut-offs with a total disregard for the poverty and inadequate income levels confronting an increasing majority' (Inner-City Community Forum 2003:190). Findings presented so far suggest asking, what exactly is the City of Johannesburg's regeneration imaginary?

> Through our regeneration [initiatives], we are going to make millionaires out of a lot of people! What is happening is that a higher calibre of people is now moving in. They are taking up the penthouses, and they are creating the world class city that we are talking about.
>
> (Interview, inner city director, 2004)

Overtly seeking a 'higher calibre of people' to transform the inner city into a 'world class' context presupposes a gentrified urban imaginary. For Slater, 'gentrification is a process directly linked to the injustice of community upheaval and working-class displacement' (2006:739). It is difficult to quantify the extent of community upheaval and resident displacement from Johannesburg's inner city neighbour-hoods, because, as sustained by Newman and Wyly, 'displaced residents have disappeared from the very places where researchers go to look for them' (2006:27).

In the neoliberal context where public policy is constructed on a quantitative evidence base, a lack of quantitative evidence regarding the number of displaced residents from the inner city results in a lack of policy to address displacement. It is, therefore, difficult to quantify the exact number of residents displaced from the 'Drill Hall', 'Turbine Hall' or the 'Bus Factory' refurbished for the purpose of creating a museum, AngloGold Ashanti's new headquarters, and a tourist attraction, respectively. As already stated, residents continue to be evicted from inner city buildings earmarked for the BBP.

> New developers want empty occupation because they can't fix a building unless we get rid of the people. For us the big issue is to decant existing tenants to other buildings, because judges often only grant eviction notices

[based on] alternative [tenant accommodation]. And that's a tough one because the City doesn't always have alternatives. I'm a great believer in market forces, and the market is profit-driven. I say to those developers wanting to make a profit: come in, we want you on board; we're trying to create a world class city. So, we need to attract the right people to live here. These are a new breed of tough developers. And the BBP is regarded as one of the most important programmes in terms of hard line regeneration interventions.

(Interview, BBP manager, 2004)

Unabashed by this 'hard line regeneration intervention', the BBP Manager speaks of a new breed of 'tough property developers' who may reinvent downtown Johannesburg via a public/private partnership with the aim of expropriating and rehabilitating buildings so that 'market forces' may prevail. This excerpt reiterates an imagined class transformation by getting 'rid of' existing residents so that the 'right people' may be attracted to live in the inner city.

At least 250 'bad buildings' have been identified for this programme, but these buildings are currently occupied by approximately 25,000 tenants (Tillim 2005). Capital investments required to rejuvenate 'bad buildings' will furthermore exclude many evictees from being able to afford rehabilitated building rentals. It could therefore be argued that as many as 25,000 tenants may be displaced from the inner city through the implementation of the BBP. Officials, nevertheless, hold on to a belief that

[f]or the inner city we want physical interventions that favour the private sector market. [As such,] we don't need social studies of the inner city. We know what the community wants. And if we are writing-off R100 million [US$14 million] worth of arrears through the BBP, this means R100 million worth of investment is going in. That's economic development!

(Interview, CoJ senior official, 2004)

Here, economic development entails melding regeneration policies with capital. Informed social studies are thus deemed counter productive as many officials claim to know what residents want. While the nullification of debts may, theoretically, be viewed as a public investment in the property market, and, as argued by Bénit-Gbaffou (2006), as a type of public subsidy for the private sector, this debt relief policy neither creates secure employment nor shared economic benefits for poor residents. Still, local politicians celebrate the fact that 'we are using international models to [facilitate] regeneration: occupancy rates are up and investments are increasing' (Councillor Cowan 2005:22). The 2030 Vision and the subsequent 2003 Inner City Regeneration Strategy undoubtedly demonstrate a preference for capital accumulation with negligible attention paid to the formulation of social policies.

Recognizing this weakness, the City responded by promulgating a Growth and Development Strategy (GDS), inclusive of a Human Development Strategy (HDS), in 2006. Both the GDS and the HDS policy documents now make a 'commitment to Johannesburg's poor [by] prioritising [resident] access to the City's social package' (CoJ 2006b:2, 4). This 'social package' entails a monthly quota of free basic public services for some inner city households. Socially responsive policies seem, simply, to be tacked onto the already established economic growth agenda as economic competitiveness continues to dominate policy discourses. Of equal significance, socially responsive policies are perceived as a means towards establishing some form of 'social cohesion' in a chaotic and transitional inner city context. Social cohesion is favoured as past governance structures, with clear divisions between public/private and economic/social roles, can no longer ensure the necessary conditions for competitive success.

> Cities in pursuit of world-class status need to strike a fine balance between economic growth and social responsibilities. To this end, social cohesion [becomes] an important resource in areas with high levels of mobility [like the inner city], and creating the means for building social cohesion is crucial to the City's goal of being a world class African city.
>
> (CoJ 2006b:10, 101, 106)

In 2007, the City Council formulated its most recent regeneration framework, namely the 'Inner City Regeneration Charter'. Through this latest framework, the city hopes to incorporate social policies derived from the GDS and HDS. However, and notwithstanding the inclusion of social policies in the 2007 Charter, the very same regeneration challenges as those identified in the 2030 Vision and in the 2003 Regeneration Strategy continue to emerge, resulting in interventions that hardly differ from those previously formulated. In fact, the Charter is specifically geared towards 'scaling up regeneration operations to ensure rapid results, [as former] City efforts have sometimes been seen as [too] localised, fragmented, and episodic' (CoJ 2007:4).

Still, the Charter does go on to stipulate that earlier initiatives 'have been critiqued [for] not [being] sensitive enough to the circumstances of poorer residents and informal businesses' (CoJ 2007:4). How economically stressed households are accommodated through this new regeneration policy and how a more sensitive regeneration approach may be facilitated by the state become particularly relevant when we consider the dire circumstances of many inner city residents: 39 per cent are formally unemployed (Leggett 2003); 62 per cent earn less than R3,500 (US$500) per month (Winkler 2006); 41 per cent pay less than R500 (US$71) a month for their accommodation (Bénit-Gbaffou 2006); and at least 10 per cent (approximately 11,200 residents) rely exclusively on the informal sector to survive (Leggett 2003).

Telling the story: evaluating Johannesburg's social regeneration policies

Social policies identified in the 2007 Charter include the municipality's 'social package', a suggestion to promote poverty alleviation and community development initiatives, and the implementation of transitional and 'inclusionary' housing projects. A deeper evaluation of each of these social policies will, shortly, demonstrate the local government's attempt to address the social responsibility deficiencies of the 2030 Vision and the 2003 Strategy; but these attempts continue to fail existing residents.

Accessing the city's 'social package' requires households to register themselves with the council as 'indigent' (CoJ 2007:55). This language is in itself problematic as many residents resent being labelled as 'indigent'. Nonetheless, once registered, inner city households are then entitled to receive a quota of free water and electricity. The city's social responsibility here, however, has less to do with supporting economically stressed households, and more to do with ensuring that 'if residents are able to access the social package this would significantly add to their ability to pay market rentals' (ibid.). Substituting municipal service payments for market rentals is perceived by the city council as a 'rent subsidy' devoid of regulations to curb escalating rentals through, for example, rent controls. Moreover, many inner city households do not hold the rates and utilities accounts for their properties as these are held by landlords. The city council then places the responsibility on landlords to register individual households as 'indigent', but this rarely happens as absentee landlordism is a common phenomenon in the inner city.

The Charter furthermore abstains from identifying specific poverty alleviation programmes with dedicated budgets. Rather, intensive urban management policies continue to be enforced with the aim of eradicating unmanaged informal trading activities, regardless of the fact that at least 11,200 inner city residents rely exclusively on the informal sector to survive.

> The current disorganized arrangement of many traders presents a key challenge to urban management. The City will, [therefore,] ensure that there is no more unmanaged trading on the streets of the Inner City beyond June 2009. Disorganized trading refers to trading without necessary permits, in an area that is not designated as a formalized trading space. A limit will be set on the number of micro-retailers that may trade in the Inner City from approved spaces. This limit will be strictly enforced [and] traders are expected to pay for the right to trade in the Inner City.
>
> (CoJ 2007:28)

Limiting micro-retailing will severely hamper employed livelihood strategies, and informal activities typically generate negligible profits rendering the city's expectation to pay for 'the right to trade in the inner city' impossible for most. Traders at designated trading spaces are also bitterly unhappy with municipal permit charges resulting in higher priced goods and fewer shoppers. And irrespective of their legal status, formalized traders do not escape police harassment (participant interview 2002). The city's formalization policy remains unrealistic in a context where the formal economy is actually informalizing, and where the informal economy absorbs those who have lost their jobs in formal enterprises (Odendaal 2005).

The City of Johannesburg has also officially abdicated its social and welfare service responsibilities, as stipulated in the legislated Regional Spatial

Development Framework (CoJ 2003b), thereby giving the local government increased leeway to abandon poor households despite the Charter's 'community development' promise. In line with neoliberal aspirations, civil society organizations are named in the Charter 'to absorb the poor through their social and welfare programmes' (CoJ 2007:36), while embracing the argument that 'regeneration achievements have been realized because private sector players took the lead and established the conditions for further private investment. The upswing in building refurbishments for middle and upper income accommodation reflects this' (ibid.:4).

The Charter does, however, make provision for '*decant* facilities to enable the relocation of residents from buildings allocated to the BBP' (ibid.:52, emphasis added). Only 10 of the 250 buildings identified for the BBP, however, will be redeveloped as 'transitional housing projects' (ibid.:53). Additionally, 'decant facilities' are deemed temporary accommodation in which tenants may reside for a maximum of two years (ibid.). Temporary accommodation simply curbs security of tenure, and the Charter remains mute on what will happen to tenants after the stipulated two-year period.

At a public meeting held in April of 2003, the executive mayor of Johannesburg stated that citizens earning less than R3,500 (US$500) per month will not be able to afford to live in the inner city (Inner-City Community Forum 2003). As a consequence, the city council will provide affordable housing for lower income earners on the urban edge (ibid.). At least 62 per cent of the inner city's current residents will, therefore, need to move as a result of 'exclusionary displacement'. This policy of displacement to the urban fringe is corroborated by an inner city ward councillor who is of the opinion that 'location does not matter for the unemployed, so they can be [displaced to] Orange Farm [on the urban fringe]' (participant interview, Bénit-Gbaffou 2006). Empirical evidence from a study conducted by the Centre on Housing Rights and Evictions (COHRE), though, demonstrates that the greater majority of residents interviewed would rather tolerate terrible living conditions than move to the urban edge as the inner city is perceived to be an easier place to survive without formal employment (COHRE 2004).

Regardless of these research findings, the Charter goes on to state that

> [t]he City does not wish to move forward on the assumption that the private sector will cater for the middle to upper income market, whereas the public and social sectors will pick up all the responsibilities for housing poorer residents, [h]owever "logical" this may appear at first glance.
>
> (CoJ 2007:53)

This 'logic' revolves around protecting the city council from 'ending up as a long-term owner and/ or manager of public housing when realistic cost recovery cannot be achieved' (CoJ 2007:53). Policies aimed at cost recovery simply 'ignite market-led growth while glossing over the socially regressive outcomes that are frequent by-products of such initiatives' (Brenner and Theodore 2005:103). The City of Johannesburg is thus squandering both an opportunity and its power to ensure the implementation of social policies that are more 'sensitive to the circumstances of poorer residents' because 'private sector providers of medium to high income housing fear that buildings housing the poor in the immediate vicinity of their developments will depress property values and prevent them from securing the right kind of tenants' (CoJ 2007:53).

Securing the 'right kind of tenants', together with a drive towards homeownership, privatization and the breakup of concentrated poverty, further incurs implementing 'social mix' policies to facilitate the desired 'social cohesion' discourse found in the HDS (CoJ 2006b) via 'inclusionary' housing programmes.

> Johannesburg will never see the problem of bad buildings addressed unless there is a huge investment in accommodation that provides a judicious mix of options for medium- to high-income earners as well as residents who are at the point in their lives where they cannot afford very much. 75,000 new residential units will be developed by 2015. Twenty thousand of these units must be affordable if the collective problem of a stressed Inner City residential environment is to be solved. This does not mean that the Inner City is to become a dormitory for the poor. The City envisages the creation of the largest mixed income community in the country, built on the basis of inclusionary housing, and the continued delivery of both medium and high-income ownership options in non-inclusionary buildings.
>
> (CoJ 2007:50)

Working on the assumption that a socially mixed community will be a socially balanced one, characterized by positive interaction between the classes, less than a third of the city's envisaged units will be earmarked for affordable rental accommodation so that medium- and high-income homeownership will prevent downtown Johannesburg from becoming a 'dormitory for the poor'. Such policy optimism, however, rarely translates into an urban context that is spatially, socially, economically and politically just; instead it leads to NIMBYism, rent increases, 'exclusionary displacement', socio-economic segregation and political isolation (Beauregard 2004; Blomley 2004; Slater 2006; Smith 2002a). Economic competitiveness, responsive governance, and social cohesion discourses, embroiled in 'renaissance' style policies, thus serve as excellent examples of how criticisms of the reality of gentrification are being deflected by different discursive policies.

Conclusion

Johannesburg's policy-makers and politicians continue to be inspired by international 'renaissance' precedents where market-led redevelopments, tax incentives, public/private partnerships, flagship projects, intensive urban management, social mix, middle- and high-income homeownership, and the disintegration of concentrated poverty all become essential regeneration strategies. In the Johannesburg case, however, an explicit policy link between inner city regeneration and economic growth is, possibly, more blatantly executed than in contexts from where these policies are imported. Such regeneration strategies usher in emblematic redevelopment undertakings while bypassing unemployment and everyday hardships regardless of the social package, poverty alleviation, community development, transitional accommodation, and inclusionary housing rhetoric found in the 2007 Charter. The City of Johannesburg's overt attempt to repopulate downtown Johannesburg with the 'right kind' of households will have a devastating impact on existing, but financially strapped, residents. Findings from this critical discourse analysis show that the 'world class African city' imaginary fails to meet socially progressive objectives, as social policies identified in the 2007 Charter are, essentially, ineffectual, and undermined by the local state, in this situated context. In short, the City of Johannesburg's current regeneration policies then stand to benefit only the new urban elite.

4 Regeneration through urban mega-projects in Riyadh

Tahar Ledraa and Nasser Abu-Anzeh

Thanks to the oil-booming economy, Riyadh city has enjoyed an outstanding rate of urban development and economic prosperity. Its area has been expanded more than one thousand times, from about 1 km² in 1950 to around 1,000 km² today. Its population has also increased over two hundred times during the last 70 years, jumping from 20,000 to 4,260,000 inhabitants (Ar-Riyadh Development Authority 2005). This soaring growth has made Riyadh one of the fastest growing cities in the world. Not every area in the city, however, has benefited from such growth and prosperity. The inner city has endured very high levels of poverty, disadvantage and urban decay. As a consequence, the city centre has suffered a massive population loss generated by the process of out-migration. At the outset, government urban policies did not directly address the issue of urban deprivation and decay. Local authorities and planning officials seem to have been more broadly focused on urban expansion and building new neighbourhoods to house the huge population influx of newcomers. Recently though, this focus has begun to shift as the decay of the urban core gets worse, the city's image is becoming blemished and many more people are enduring severe deprivation.

Riyadh inner city regeneration process

The inner city of Riyadh was once the liveliest retail and administrative centre of the city. It harboured important commercial uses and government institutions. It was also the main cultural and religious centre containing the main mosque and public gathering places where popular festivities and official ceremonies were held.

The urban pattern was an organic structure with curvilinear narrow streets and muddy building constructions in a pedestrian-oriented design layout. With the booming economy and the introduction of the automobile, new gridiron car–oriented planning layout and concrete structures were adopted to characterize the newly developed suburban neighbourhoods. Many local residents flew the inner city and its dense traditional layout to settle in these newly planned areas with larger modern housing and wider street patterns.

As a result of population out-migration and abandonment, the inner city suffered severe degradation and decay. After a long neglect, local authorities decided finally to tackle the problem of its decline. A three-phase regeneration plan was set up to revitalize the area. The first phase consisted of rebuilding and expanding the Government Compound which comprises the regional governorate, city hall, state mosque and public parks and gardens with some big shopping centres around them. The second phase was concerned with cultural flagship development of the King Abdulaziz Historical Centre (KAHC) containing the national museum, the conference hall, and public library. The third stage, which is currently underway, is devised to regenerate the area in between, that is, the Dhaheera quarter and is supposed to cater for high-quality shopping and services. It also links the administrative centre to the cultural quarter. The objective was to reinstate the same old functions of the inner city, but in a highly brand new modern style, and revitalize the area once again through recalling its functions and re-branding its image. The following sections will have a closer look at each of these phases.

Rebuilding and expanding the Government Compound

From the outset of the regeneration process, it was decided to renovate and expand the locus of power and the main institutional buildings symbolising the local authority and control. The objective was twofold. First is that Riyadh, as the centre of power and the capital city of the state, merits highly modern building structures with wide streets and large open spaces to exhibit ostentatiously to the world. Second, the revitalization of the surrounding areas for commercial activities and services would follow up to cater for the masses of visitors that would be drawn to the area for their administrative and other services. The argument was that renovating and expanding the government compound would lead to reshaping the inner city as a highly modern complex with an extravagant planning layout and design which would not only help hold back the process of inner city decline, but would also assist in engaging the process of its regeneration.

To achieve that goal, the Government Compound had to be extended over an area of 11,500 m² to harbour new official buildings, that is, the governorate, the city hall and the central police headquarters. The main state mosque, as it embodies the religious institution, a pillar of local power, was placed adjacent to the governorate palace and covered an area of 16,800 m² (see Figure 4.1). For the whole complex to convey a sense of an impressive monumental image and an imposing spatial order, a network of vast open spaces were created with large arterial roads leading to and from the compound. In quantitative terms, this has meant that an area of 71,380 m² needed to be cleared in order to make room for the whole complex to be built – over 590 dwelling units were deemed immaterial.

Since the area was re-zoned for administrative uses where strictly official buildings were allowed, local authorities did not face serious difficulties in clearing the land. When it comes to expropriating property owners for reasons of establishing power institutions, the authorities usually lean on a system of sticks and carrots to go ahead with their plans. Owners whose properties were affected by the compound expansion were compensated. When it comes down to paying back for land expropriations, the authorities would not be sparing in the use of public money. They can also refer, if required, to the regulation that allows them to acquire land in the public interest. After all, it is not very common, for a society that is not accustomed to democratic opposition, to stand against the will of its powerful leaders.

If local public authorities have taken for themselves the task to rebuild the whole administrative zone to make provision for government buildings and institutions, private investors were called upon to take on the follow-up process of developing the remaining commercial zones within the complex. The logic behind the local planners' initiative was that the renovation of the administrative centre would in turn generate the type of spin-off activities that would add to the liveliness of the urban core as a whole. For that reason, public money was used to pay for all the required services and infrastructures within the regenerated area to attract prospective investors. Private investments, however, have been limited only to the use of the lots bordering arterial roads, leaving the whole area behind to face its fate. In other words, this has created a situation where only commercial strips were developed hiding a whole derelict area at the rear.

But here, the process of clearing land and slums for commercial developments was not as smooth as it was in the case of the government compound. Since the area was re-zoned mainly as commercial, many private interests were at stake. The market rules and land speculation were left to play their role. Local authorities did not want to interfere to compel owners to property expropriation against their will. Although many of them have agreed to sell, some others have been quite reluctant. Many obstacles have thus ambushed the commercial shopping centres projected to be built.

Since the area was populated well before the modern bureaucratic administration was established in Riyadh, many property owners do not hold official papers to verify their claim, though their ancestors had been living there for ages. Being the oldest part of the city, many other properties have fallen into ruin and it became difficult to determine the limits. A lot of properties were also interpenetrating as a result of successive partitions among inheritors. Determining a property right was not that easy, let alone compensating the owners. On top of this, the market speculations over land prices have rendered the negotiations process a very daunting business. Al-Muaygaliyah shopping centre, for example, is a case in point. It took 25 years for the problem of land to be settled down and the shopping centre to finally be built.

The lengthy process of resolving the issue of the land has put off many investors from getting involved in this development. The Ar-Riyadh Development Authority (ADA), the development arm of local public authorities, decided then to create a joint venture company where the municipality holds the majority of the shares (50 per cent) to help establish confidence among private investors about the project.

In order to resolve the issue of land and slum clearance, the ADA was vested with decisive powers and substantial annual funding enabling the physical regeneration of buildings and land in specific areas. It had the power to grant planning permission for projects within the designated area, in addition to the power of compulsory purchase that allows them to acquire, hold, manage and dispose of land at their discretion.

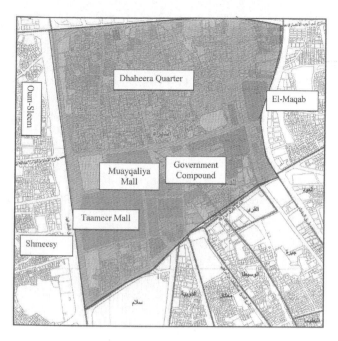

Figure 4.1 Map of inner city Riyadh, showing Dhaheera Quarter

Source: © Ar-Riyadh Development Authority.

Despite being very unpopular, hard interventions through gentrification policies were adopted. The state-sponsored gentrification strategy was retained to show the authorities strong commitment to the regeneration process. By such commitment, two objectives were meant to be reached. The first was to encourage private sector investors to be part of the process. The second was to impose on the city the image and the symbolism that would express and reflect what Riyadh ought to be and look like.

Central to this re-imaging has been the development of a major leisure and cultural complex, the KAHC. This complex constitutes the bulk of the second phase of Riyadh's inner city regeneration. The following section is devoted to explain it.

The King Abdulaziz Historical Centre or culture-led regeneration

The area of the KAHC is defined as extending to some 37.5 hectares and located just to the north of the city's administrative compound and retail and commercial core. The area is roughly rectangular in shape, and is bounded by main arterial roads from all sides. The centre comprises public open spaces, a museum complex and many other places for cultural and leisure activities. The ADA was given only 32 months to get the whole development ready for the centenary celebrations commemorating the state foundation which coincided with the year 1999.

Particular imaginations of the city's history have been drawn upon to link the development to particular meanings and understandings of place. Many traditional buildings were maintained and restored which by their symbolism and meanings were intended to create new attachments and make the place more marketable and legitimate.

The development works were largely managed and administered by ADA. This state-owned company was engaged in acquiring, renewing and reassembling properties to resell them as larger lots to investors. It has also undertaken development schemes on its own volition or as joint ventures with private owners and developers.

The contention was that, as the earlier phase of regeneration has mainly focused on the renovation of the administrative and commercial core, the inner city still suffers from the lack of one major dimension, that of a strong cultural core devoted to popular festivities and leisure activities. The intensity of such cultural activities would lead to the creation of lively public places and parks. In this case, not only would the inner city get rid of the negative images associated with the older squalid deteriorated areas, but would also be elevated to its position as a new postmodern, consumption-oriented inner city. It would, hence, become a visitor destination offering good quality of life for professionals, higher income and skilled workers, which would result in improving the overall social and economic life throughout the whole inner city.

As was the case for the administrative area, the central government took on its shoulders the burden of financing the whole cultural complex from beginning to end. The total costs amounted up to US$ 186 million. When the government decides to go ahead with a project for its own institutions, no major hurdles are laid down. Property expropriation was not a serious issue in the case of KAHC.

Since the intention was to design spaces for cultural projects in the form of prestige art events or flagship developments, the KAHC area was delineated as 'planning free'. The standard development control regulations were made null and void, creating an atmosphere for aesthetic creativity which is precisely what architect designers look for. Yet, in their quest for city image beautification, official planners have fallen short of giving as much weight or concern to the underprivileged tenants who populated the area.

It must be mentioned that property owners do not very often live in the inner city. They have already left to settle somewhere in the new suburban neighbourhoods. The majority of the dwelling units are either rented or unfit for human occupation as they suffer high levels of degradation. This is the reason why local authorities have not faced much resistance in clearing the land for KAHC redevelopment. With regard to tenants, they were forced to seek refuge in other neighbouring derelict areas where they can still find relatively lower rents. Many of them decided to settle just nearby in the Dhaheera quarter, the third phase of the regeneration process which is explained in the following section.

The regeneration of the Dhaheera quarter

The Dhaheera quarter is a derelict area situated in between the government compound and KAHC and extending over 7.5 hectares. The area was supposed to benefit from the spin-off effects of the previously regenerated areas. It has been waiting for developers to come to revitalize it for more than a decade now. Finally, local authorities have decided to take on the redevelopment works in the hope that private developers will follow suit and get involved in the process. The adopted masterplan for the area indicates that it would be essentially re-zoned for commercial and office uses with some residential flats on higher floors.

The project is currently underway. The idea behind the regeneration of the Dhaheera quarter is to have, once completed, a strong vibrant inner city containing administrative, religious, cultural, commercial and residential uses assembled within the contours of one whole renovated area. It would ultimately convey a powerful positive city image that would in turn lead to a rejection of older tarnished images of the city centre.

Judging from the orientations of the area's masterplan (ADA 2004), the regeneration approach adopted is mainly middle-class consumption oriented. Only stores providing the highest quality range of shops, leisure, and entertainment facilities for relatively well-heeled citizens are to be developed. The disadvantaged people who have previously inhabited the area have not been taken into consideration. The high brand shops of the latest fashionable items are simply beyond the reach of these less well-off populations.

Urban regeneration and community involvement

The regeneration process as undertaken in the case of Riyadh has not been bound up with participation and involvement of the community. The little success so far of the regeneration projects in Riyadh's inner city may be attributed to a great extent to the lack of involvement among the local community. The new commercial developments for example, namely Muayqaliya and Taameer shopping stores, were unable to draw larger numbers of shoppers simply because those who live within their market catchments area, that is, the young, the elderly, the socially excluded and those from ethnic minority backgrounds, have not been targeted by these stores.

The KAHC, with its extravagant modern buildings and spaces, is very much seen as an elitist development. The rush of the local population to take advantage of its facilities simply has not happened. Many locals do not feel they are concerned by the project and, as such, are not inclined to use its facilities. Even the open public spaces provided in this cultural quarter have not drawn the masses of people to use them (see Figure 4.2). What is striking, however, is that although young people and teenagers make up the majority of the city's population – 55 per cent below 30 years (ADA 2005) – they are not really keen on using these spaces. Needless to say that they were not included in the planning and their needs and concerns have not been central to the regeneration. Instead, the focus has been on designing new public spaces from which such groups, perceived as a source of disarray and threat to security, are monitored and excluded.

If the substantial public funds poured into the inner city regeneration process have succeeded in re-branding Riyadh's image as a centre for high-quality consumption, they have failed to add any significant difference to the lives of the

Figure 4.2 Urban vitality in central Riyadh. Inner city people prefer being in these lively, crowded, dilapidated areas rather than in the well-managed spaces of the regenerated cultural quarter.

Photograph by Tahar Ledraa and Nasser Abu-Anzeh.

82 per cent of its residents who hold tenancy rights there. As long as the identities of local communities are obfuscated and the needs of less well-off consumers are not considered, the sustainability of the project faces serious questions in the near future, particularly at times of economic recession. The involvement of local people and communities should have been formally instilled within government guidelines for regeneration policies.

The impacts of urban regeneration

The authorities' approach to tackle the issue of inner city regeneration was completely at odds with the views and expectations of the local population. Since the main issue in the eyes of local officials was one of physical dereliction and image degradation, the approach was one of inner city beautification and flagship development. Local residents, however, see the problem as a shortage of job opportunities, adequate affordable housing and services. As such, they would expect actions and programmes that would help them get out of their poverty trap. This difference in approaches has had some devastating impacts on the area and the people who lived there.

In fact, many residents were forced to endure residential displacement. The categories that suffered most from such dislodgement were those at the bottom of the social ladder. The ADA statistics indicate that over 82 per cent of families hold only tenancy rights over their dwellings. This explains a great deal of their vulnerability for eviction and displacement. Not only has this relocation imposed hardship on poorer families, but made them move away from their sources of employment and livelihood.

As long as the ultimate aim for ADA officials was to attract people, investments and jobs back in the revitalized central area, the people targeted were thus outsiders to the area. There is, therefore, a mismatch between the people who endure the poverty problem and the target groups for whom these regeneration strategies are designed.

The area under gentrification was quite large, and it has triggered some profound social-demographic changes, to the point where life became unbearable to some social groups whose lives were quite shattered when their social networks were destroyed. This urban gentrification has torn apart their set of established social connections and many residents have found it difficult to cope with the new imposed social setting. Although it looks decayed, the old urban pattern holds some strong cosy social networks and is regarded as a homely place for many people (see Figure 4.2), particularly the most vulnerable, the elderly.

The other impact is the emergence of some problematic areas with concentrated marginal groups as a result of selective mobility among social groups. Less affluent households were constrained in their residential choice. As the supply of affordable units shrink, the demand gets higher and the competition becomes fierce within the bottom housing submarket. While the areas concerned by regeneration interventions, the Deera and Fota quarters, have seen their residential densities drop to 145 and 148 persons per hectare, respectively, poorer neighbouring ones have seen their densities jump to 386 persons per hectare for El-Marqab and 337 persons per hectare for Thleem (ADA 2005). One might conclude that inner city regeneration as applied in Riyadh did not solve the problem of overcrowding and urban decay but simply displaced it to adjacent dilapidated areas to render the situation even worse.

Market-oriented land redevelopment has introduced a profound change in the land uses in areas under regeneration. The old uses that once offered the kind of activities where unskilled and disadvantaged people were absorbed were bound to disappear to make room for more competitive uses. Profit-making projects such as retail businesses, shopping centres and offices were built in their place.

Riyadh's inner city regeneration has not only disrupted the socio-economic life of the people and their activities but affected the local environment as well. Pollution and congestion have been exacerbated by recent developments to a point where Riyadh centre now has some of the worst air quality and traffic conditions than any part of the city. As the interest for image beautification and profit making dominated the regeneration agenda, issues related to environment, sustainability and social inclusion have been marginalized.

Conclusion

In summary, it is now clear that urban regeneration as applied to Riyadh's inner city has imposed hardship on disadvantaged residents who were forced to move. Their displacement has led to a further deterioration of dilapidated adjacent quarters. The problem with Riyadh's urban regeneration is that it has lacked a clear vision and a comprehensive view that embraces the blighted area as one whole, together with the socio-economic conditions of the disadvantaged people who live there. This example shows the limited success of such type of urban regeneration that is place centred, not people focused. The project-led approach to regeneration has proven to fall short of dealing with the multifaceted problem of urban decay, economic downturn and social deprivation.

It would have been a lot more beneficial for the area and the local people if the patterns of mixed land ownership were not damaged, so that self-improvement and small-scale investment in property would have been possible. Policy-makers ought to have adopted an overall approach to property management and upgrading based on balancing the need to improve the area's environment with the need to retain existing activity. In fact, the availability of differing unit sizes of property at varying degrees of cost would have allowed small businesses to gain a foothold and not be driven out of business by sudden rises in rent or property values. The experience of Riyadh demonstrates that bigger lots and larger businesses do not necessarily bring about vibrant urban life and active street frontages.

The impact of the KAHC massive cultural structures with the festivals and events they support remains very insignificant on both the liveliness of the area and on the disadvantaged people who were forced to relocate to free the site for these prestigious symbolic projects. The type of people who are attracted to these structures are those of middle and upper social classes for whom the inner city is not a place of residence. Through these projects, the regenerated area shows a clear cultural bias in favour of these social strata and their modes of consumption.

The case of Riyadh shows that the pressure to re-brand the city image through flagship development and high-return market-driven programmes has the potential to undermine more socially oriented agendas. Development projects are deliberately generated to promote profitability and aesthetics at the expense of excluded communities and disadvantaged residents. Affordable housing and job opportunities for the unskilled have not been an area of concern in the regeneration plans. The private sector developers seem to be unwilling to embark on schemes which, in the short- to medium-term, will yield fewer market returns. Housing development in the regenerated area has lagged behind other more profitable commercial developments. As costs of living have rapidly increased, more and more low-paid workers and even those on moderate incomes face deeper exclusion from circuits of consumption, such as housing. In the absence of strong redistributive planning frameworks, the inequalities and exclusion perpetuated by this type of regeneration policies may grow even further.

5 Regulation and property speculation in the centre of Mexico City

Beatriz García-Peralta and Melanie Lombard

Living in Mexico City involves coexisting with its rich cultural and historical heritage, but also with social inequality, lack of safety on the streets and the hopes of the many ordinary people who built the city, for whom access to housing is a principal concern. More than half of Mexico City's housing is self-built (García-Peralta 2005), and the importance of providing decent, affordable housing for the low-income majority cannot be overemphasized. Mexico City's urban sprawl, which is well documented, and the depopulation of the city centre, which is less so, are the result of intermittent crises which worsened during the 1980s, a decade that saw trade liberalization and financial deregulation. These events led to the deterioration of working-class purchasing power. But the continued need for affordable land and housing for the majority of residents in Mexico City also reflects the lack of urban policy addressing these issues.

This chapter looks at how in 2000, an elected metropolitan authority in Mexico City tried specifically to address the issue of sustainable urban development, with the by-law known as *Bando Dos*. The term 'by-law' is used here to describe this administrative order issued by the Federal District Government, which is more general and legally weaker than a regulation. Passed by the Gobierno del Distrito Federal (GDF or Federal District Government), its aim was to promote the regeneration of the city centre and access to affordable housing (GDF 2000). This by-law had three primary objectives: first, it sought to restrict the growth of housing and commercial developments in designated environmental conservation areas; second, it attempted to reverse the depopulation trend that had been observed over the past 30 years in the central boroughs; and third, it endeavoured to recover space and guarantee the supply of serviced, well-located land to build housing for the impoverished population. However, the legal instruments necessary to fulfil this last objective were not established. What was needed was a regulation prioritizing the collective interest and restricting the effects of the open land market, namely the exclusion of social and public uses in favour of income-seeking developments in the central boroughs. The by-law elicited great controversy because of its implications, which are further discussed in the following sections, after a brief look at some of the characteristics of Mexico City and its urban policy.

Urban policy challenges in Mexico City

Mexico City experienced rapid and dramatic economic and population growth between 1950 and 1980 when the population grew from 2.9 million to 13 million inhabitants. As a result, by 1995, it had become the world's second most highly populated city (Garza Villarreal 2000:229). This was a rather dubious honour as Garza Villarreal (2006) so rightly points out, since in a comparative ranking of 66 metropolitan regions, according to real GDP per capita and competitiveness levels, it was barely in sixty-third place. The socio-economic disparities of this megacity, now home to 19.2 million people, pose enormous economic, social, political and urban challenges.

These complex socio-economic conditions are compounded by the fact that the city has only had an elected government since 1997. Mexico City's democratic reactivation has sparked conflicts between the leaders of the various political parties in the political and administrative units (boroughs and municipalities) comprising the city. Thus, the city government, ruled by the leftist party since 1997, has to negotiate with a national president and a state governor from different political parties. This situation hinders the implementation of urban policy, for which the provision of suitable housing for the masses constitutes one of the greatest hurdles.

Favourable economic conditions during the 1960s and 1970s led to the emergence of a middle-class, and the creation of the institutions that have financed the majority of social housing in Mexico. The Fondo de Operación y Descuento Bancario a la Vivienda (Fovi or the Fund for Bank Operations and Housing Funds) was set up in 1963, based on the obligation imposed on banking institutions to allocate part of their revenues from deposits, in order to grant loans for the acquisition and/or construction of housing. In 1972, a compulsory quota of five per cent of workers' salaries was established to create revolving funds that would provide resources for the construction of housing for these workers. The Instituto del Fondo Nacional de la Vivienda para los Trabajadores (Infonavit or National Funding Institute for Workers' Housing) became the most important housing fund since it had the largest number of affiliated members (private sector workers). The creation of housing funds exempted employers from the obligation to provide housing for their workers, originally enshrined in the 1917 Constitution, whereby owners of firms with over 100 employees were obliged to provide those employees with 'comfortable, hygienic accommodation' (García-Peralta 2005:1).

It is undeniable that social housing in Mexico has never received sufficient fiscal resources. The Federal Government's main role has been establishing the legal and technical bases as well as administering the funding originating from savers and workers to finance the production of social housing. Thus, social housing has become an important sub-market for the business sector concerned with constructing and marketing housing. What has particularly characterized the housing market in Mexico is the lack of investment by the private sector and a lack of social rented and affordable housing for the lower-income sectors, despite the fact that the most important financial resource for housing is the contribution of workers who make up such a large proportion of the population of Mexico City, and the country as a whole.

Table 5.1 Population in Mexico City Centre, 1970–2005

Borough	1970	1980	1990	1995	2000	2005
Benito Juárez	576,475	480,741	407,811	369,956	360,478	355,017
Cuauhtémoc	923,182	734,277	595,960	540,382	516,255	521,348
Miguel Hidalgo	605,560	501,334	406,868	364,396	352,640	353,534
Venustiano Carranza	749,483	634,340	519,628	485,623	462,806	447,459
City Centre	2,854,700	2,350,692	1,930,267	1,760,357	1,692,179	1,677,358

Source: Based on data from Population and Housing Censuses (Instituto Nacional de Estadística Geografía e Informática (INEGI 1970, 1980, 1990, 1995, 2000) and Second Population and Housing Survey (INEGI 2005).

From the 1990s onwards, social housing in Mexico has been regarded primarily as a financial business rather than a social good or a means of alleviating the low-income housing crisis. Despite government rhetoric restricting housing institutions to acting within a purely financial role, their participation included actually facilitating the opening up of the market to the financial sector and foreign firms. This confirms that the role of the state has been more responsive to the needs of the various private actors than to those of the poorest sectors of society.

Bando Dos: building the city of hope or ...

During the last three decades, the Mexico City inner city area suffered depopulation. The four central boroughs lost 40 per cent of their population (see Table 5.1), leading to many problems including urban decline, under-use of infrastructure, peripheral growth and the saturation of an inefficient metropolitan public transport system.

Particularly since 1997, attempts have been made to establish various policies and instruments to reverse the phenomenon of depopulation in the four central boroughs, such as the 1997 General Urban Development Programme for the Federal District (Asamblea Legislativa del Distrito Federal 1996), and Borough Urban Development Programmes and Partial Programmes (Secretaría de Desarrollo Urbano y Vivienda 2001). Housing is a crucial element in achieving the urban regeneration of the city centre, as keeping an area safe and in good condition obviously requires a mix of residential and other uses, to prevent it from becoming a no-man's land at night.

In this context, it is important to note that the by-law issued by the city mayor in 2000 became the administration's most important urban policy instrument, even more than the Urban Development Programmes, guiding the Federal District's real estate activity. Bando Dos also included other aspects such as the protection of the water table, but these will not be explored here, since this chapter focuses on housing.

Despite its 'legal weakness' in comparison with other legal regulations, Bando Dos provided certainty for real estate developers regarding the feasibility of service provision and hence the guarantee of obtaining a building permit in the four boroughs. Previously, when a plot of land was purchased to build housing, there was no certainty that this permission would be obtained in order to develop it. The strength of Bando Dos lay in its denial of permission for service provision (principally water) for the construction of housing developments outside the central boroughs, while expediting this permission within the four central boroughs. Due to the urban diversity of this zone, the measure benefited the real estate sector operating in higher income neighbourhoods. This created a boom in demand for land solely in the four boroughs with conditions of economic stability. In the light of housing need, unmet since the 1994 economic crisis, this policy outcome seemed to fly in the face of the regulation's objectives. Owing to the implementation of Bando Dos, more than 80 per cent of this land ended up in the hands of 14 estate agents, leading to an increase in land prices as well as in the average price of land for social housing apartment blocks, which rose from US$3,500 to 9,000 per unit over a period of six years from 2000 to 2006.

The by-law did not impose any restrictions on land speculation or on the type of housing that could be built. Consequently, both landowners and developers seized the opportunity to benefit from this. Nor was there any coordination between the Federal District Government and the Government of the State of Mexico in anticipation of the repercussions the measure might have on the municipalities of the Mexico City Metropolitan Area (MCMA).

The data allowing the evaluation of the impact of this policy are taken from the 2006 survey by Desarrolladora Metropolitana (DeMet, a development company based in Mexico City), which since 1999 has conducted a quarterly survey of all the new private real estate developments in the MCMA. According to this study, Bando Dos effectively shifted the supply of housing in the Federal District back to the central boroughs. In the year 2000, these boroughs accounted for 30 per cent of the housing supply, a figure that had risen to 72 per cent by 2005. During the period that the by-law was in force, private developers sold 37,800 new private dwellings in these four boroughs. This means that nearly 135,000 inhabitants have remained in or returned to this part of the city as a result of Bando Dos, in addition to the efforts made by the city government through Instituto de Vivienda del Distrito Federal (or Federal District Housing Institute), which constructed 2,360 social housing units (mainly apartments) in the central boroughs (Tamayo 2007).

However, private developers produced largely middle-income housing, which contributed to the expulsion of the working class towards the city periphery. According to Infonavit data, during the period from 1996 to 2000, of the 84,382 housing credits granted to workers registered in the Federal District, 57 per cent of them were used to purchase dwellings located in the suburban municipalities of the State of Mexico. The urban sprawl of the MCMA comprises 16 boroughs in the inner city Federal District, 59 municipalities in the State of Mexico and one in the State of Hidalgo (see Figure 5.1). From 2001 to 2005, this institute granted 122,317 credits to workers registered in the Federal District, 71 per cent of whom purchased homes in the State of Mexico, a figure that had risen to 80 per cent by 2005. At the same time, established zones of low-income neighbourhoods in the periphery have become increasingly overcrowded. Whilst there are unfortunately no statistics to provide a detailed picture of this phenomenon, it is widely known that many of the families that own homes in these neighbourhoods build rooms to rent out.

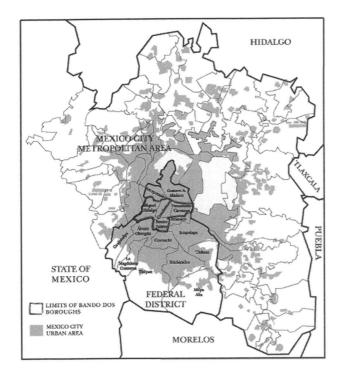

Figure 5.1 Map of Mexico City metropolitan area.

Source: Instituto Nacional de Estadistica Geograffa e Informatica INEGI wwww.inegi.gob.mx (National Institute of Statistical Geography and Computer Science.

... or the city for real estate capital?

Based on the above, it is quite clear that the beneficiaries of Bando Dos were the landowners, given the 150 per cent rise in land prices, and housing developers, who took advantage of the ease of obtaining permits in the four boroughs and undoubtedly made enormous profits. The density of housing developments rose from 350 to 650 housing units per hectare, and the size of the different types of dwellings decreased while prices rose; in the Cuauhtémoc borough, the cheapest new housing units, apartments of 34.7 square metres designed to accommodate low-income households, were sold at a higher price per square metre than certain older luxury residential flats in high-income neighbourhoods. The small amount of social housing now available in these boroughs is located in areas with proper infrastructure and services, but in developments with twice the density they had in 2000, and in smaller, more expensive apartments (Benlliure 2006).

Moreover, the fact that policies are not properly implemented, and the lack of a shared metropolitan vision on the part of local authorities, leads to unwanted effects such as the growth of the periphery. The scope of the phenomenon that took place in the municipalities comprising the MCMA is apparent: between 1999 and 2005, 276,197 dwellings were authorized for 1,243,246 inhabitants. This irrational growth, along with the privatization of social housing in the city, has led to the loss of 3.2 million eight-hour working days in 2005 through working hours spent in transit, as peripheral developments often suffer from lack of adequate infrastructure, compounding their distance from the rest of the city (García-Peralta 2006).

This account shows that the city government does not realize that its stated aim of reorienting growth and promoting the construction of housing for the low-income population in order to regenerate the city centre 'for everyone' – in other words, with an inclusive vision – requires handling the housing problem differently. Two issues are critical: first, due to families' low-income levels and the informal work conditions of the majority of the population, it is impossible to solve the problems of housing and urban regeneration through private market mechanisms or even in a public sub-market similar to the existing one. Second, it is important to open up a discussion around the private ownership of social housing. The privatization of social housing, which in Mexico has been justified as a way of guaranteeing property ownership as a patrimony for economically and socially weaker households, has become a perverse instrument of price increases, spatial segregation and the decline of the central city. It therefore fails to help the city centre recover its social fabric by retaining and attracting people from various social strata. Thus, although the leftist government's rhetoric stresses social benefits, what actually prevails is the ideology of housing privatization imposed in the 1960s and reinforced by the World Bank in the 1990s, in order to facilitate market operation. This limits the scope of inclusive urban regeneration, and distorts the reorientation of growth.

Although they have failed to meet adequately the population's housing needs, housing policies, including Bando Dos, have contributed to the development of the private construction and finance sectors. Given the benefits for developers arising from the by-law, as borne out by the current housing supply, they propose to continue enjoying the privilege of low-risk, rapid procedures allowing them to continue their real estate business unfettered in the rest of the city.

The Sorcerer's Apprentice

The construction of housing promoted by Bando Dos caused great discontent among the inhabitants of the central boroughs, who had a distorted vision of the impact of repopulating the zone, given that the infrastructure in these areas was underutilized. In fact the by-law had the opposite effect to that expected as it increased the cost of land even in the lowest-income neighbourhoods of these boroughs. In order to genuinely provide housing for the lower-income sectors, increased state intervention in the land market is needed.

Conclusive evidence of the impossibility of providing housing in this way for the lower-income sectors in the MCMA can be found in the statistics on supply and demand of housing. From 2000 to 2005, 153,886 housing units were required for households whose main breadwinner earned less than two times the minimum wage. Private developers did not offer any housing for this sector, while for households earning more than ten times the minimum wage, there was an over-supply, a situation which appears not to concern the real estate investors, as long as prices remain stable (Benlliure 2007).

Ultimately, Bando Dos had the effect of multiplying the upper end of the real estate market. This brings to mind the story of the Sorcerer's Apprentice, where the actions of the Federal District Government recall the lines in which the apprentice implores the returning sorcerer to help him with the mess he has created. By now a cliché, 'The spirits that I called', a garbled version of one of Goethe's lines is often used to describe a situation where somebody summons help or uses allies that he cannot control, especially in politics or in this case the land market. In this case, the sorcerer should have been represented by a team of expert planners and economists which the government should have called upon for advice on its proposals for this complex city. However, the missing element from this story is a consistent and integrated planning vision. In order to promote policies such as Bando Dos, aimed at preserving the public interest, the spatial planning policy for the city as a whole must integrate instruments with the object of guaranteeing a more productive, equitable, socially and environmentally sustainable city. This includes preparing measures to prevent practices where the outcome, far from contributing to the right to housing as enshrined in the Constitution, actually limits access to social housing. This story shows that in attempting to offer inclusive residential space in the centre of Mexico City, the authors of urban policy must not be taken by surprise by the effects of their own measures. There are examples of land price regulation, expropriation and limits on speculation which should be applied in metropolitan frameworks. However, these measures may only be possible in the event of a change in the economic model currently in place in Mexico.

Acknowledgements: We would like to thank the architect Pablo Benlliure, housing professional and leading market expert, who supervises the DeMet survey, for providing access to this information.

6 Museumization and transformation in Florence

Laura Colini, Anna Lisa Pecoriello, Lorenzo Tripodi and Iacopo Zetti

Usually, the term renaissance is not used to describe Florence's regeneration programs, probably because there is no other possible Renaissance but the original in the Florentine vision. All over the world, Florence recalls an image of romanticism, a special place where, in the fifteenth century, human nature developed some of the highest expressions of creativity which would influence the arts and knowledge of generations to come. Hence, any possible future scenario for the city has to deal with its strong global identity of an iconic historic location due to its unique and precious cultural heritage. It is a controversial aspect, as it can be both an incentive and a limit in the face of modernization and urban transformations.

This condition strongly influenced the urban life, due also to the typical Florentine attitude, characterized by a polemic esprit, if not a real quarrelsomeness, pervading the political as well as the everyday life sphere, and transforming every planning process into a complex and conflicting argument. On the other hand, the myth of a public's contentious attitude has been regularly used as an alibi to foster decisions overriding consensual processes.

This chapter will outline general tendencies transforming Florence and in particular the historical centre, the effects of which are questioning the right to live, access and inhabit the whole city. It will look at the rhetoric deployed within the plans to sustain the regeneration programmes and the contradictions between assumed goals and proposed solutions. It will focus on the role of infrastructural programmes and related Public Private Partnerships (PPPs) and eventually on the case of Piazza Ghiberti as an example of redevelopment processes reshaping the city centre.

Regeneration programmes and the city centre

In the last years, Florence witnessed an intense rebirth of urban regeneration activities, with new building sites mushrooming all over the city, accompanied by loud promotional campaigns advertising a vision of high-quality city living. The regeneration plans are characterized by large-scale construction programmes that often correspond to the re-actualization of long-debated infrastructural and residential projects. They are developed in the frame of public policies that reinforce the uncontested dominance of tourism and other forms of exploitation of rent while expelling most elements of diversity and cultural innovation from the city.

Such policies endorse and accelerate a process of alienation of the historical centre from the whole city. The inner city is becoming an exclusive leisure district in a network of globally valuable tourist locations, exploiting consistent economic flows derived from a consolidated historical image (Tripodi 2004). Florence municipality experienced a significant loss of 11.7 per cent of residents between 1991 and 2001 (Italian National Institute of Statistics 2001) and in particular the active section of the population is moving to satellite municipalities.

Such dynamics intensify commuting back to the city, alongside the already significant influx of tourists, and affect the development of the entire metropolitan system growing around the pivotal historical core. The redesign of the city centre expels residential life as well as traditional functions, displaced by market-driven pressures such as the increase of real estate values in central locations, accessible only for profitable activities or temporary users; the actual need for more suitable spaces that cannot be provided by historical buildings, maintenance costs of which are very high; the dominance of the tourism economy which undermines existing facilities for residents in favour of services dedicated to temporary users; and the increasing difficulty of accessing the centre for private car owners and daily workers who cannot rely on the public transport system.

Change of demographic profile, lack of access to affordable housing, impoverishment of cultural life and standardization of commerce are some manifest consequences. Long-time residents are moving to the outskirts where they can benefit from new housing and facilities, avoiding some of the distressing tensions in the historical centre such as dysfunctional mobility, swarms of tourists colonizing public space, cultural clashes with newcomers, and the diffuse perception of urban degradation encouraged by anxiety-inducing media campaigns. In the meantime, low- or precarious-income inhabitants live in the lowest quality buildings still existing in the historical area. They are mainly single-family households (often seniors) or new migrants living crammed into overpriced apartments. Italian students, once a consistent part of the city centre's population, are moving towards the periphery in search of more affordable locations closer to new university campuses. They are replaced by an increasing population of foreign students from numerous international universities, art, fashion design and language schools, recently developed in the centre. Alongside the mass of tourists, estimated at approximately six million per year (Comune di Firenze 2006), the movement of international, educated people feeds the image of the centre as a golden spot for investment, tourism and temporary residence (see Figure 6.1).

The whole central district is turning into a gentrified urban island. This is a peculiar form of gentrification because the lower income population is not replaced by the upper-class but by the steady pressure of temporary users who are ready to pay any rent for short-term stays in Florence. Supported by the mighty pressure of real estate investments and by wealthy citizens' craving for easy profits, residential estates are often refurbished into smaller units to let or used as bed and breakfasts. Many buildings, once hosting collective functions, are being turned into hotels. Traditional meeting places such as the leftist Casa del Popolo (House of People) and other workers' movement facilities are changing and losing their social role.

Figure 6.1 Florence city centre: consuming the image of the city.

Photograph by G. Pizziolo.

Much of the traditional commercial fabric such as grocery stores, open air markets, and art and crafts activities are suffering the high cost of renting spaces and are slowly disappearing. A similar fate is also affecting old-time pharmacies, bookshops, cinemas and literary cafes, replaced by luxury shops clustering in monocultural districts dedicated to expensive fashion or retail chains and franchise shops.

In the meantime, cheap shops often run by migrant labour are flourishing all over the city. Many of them sell low-price, Florentine-style wares, imitating while trivializing the traditional image and quality of the local production. They effectively respond to tourist demand for consumption of souvenirs, clothes and accessories. However, these ethnic shops also represent a resource supplying cheaper basic goods such as groceries.

On the other hand, residents inhabiting Florence's inner city have organized themselves in grass-roots groups. Around 40 citizen committees now form an umbrella organization, firmly reclaiming a voice in the debate about specific urban issues and the general philosophy that informs the city administration's choices and management. Citizens contest the regeneration programmes claiming that they cause environmental problems, social injustice, discrimination and a commodification of urban life. The public administration is often blamed for designing and implementing public policies in splendid isolation, inconsiderate of the welfare of their citizens.

Planning instruments and their rhetoric

The vision for the urban regeneration of Florence is presented in two planning documents: the Strategic Plan and the Structural Plan. The Strategic Plan is not a legally binding land-use plan, but following a tendency in contemporary planning practice, it represents a concerted vision for future urban development (Sartorio 2005). It involves a large partnership of private and public stakeholders and representatives of neighbouring municipalities, chaired by the Mayor of Florence.

Florence is envisioned as a cultural centre for branded Italian production and for high-quality handcraft that encourages and manages tourism, promoting a new image related to creativity and technological innovation. The main objectives resulting from a negotiation among the most powerful players are identified as follows:

- 'Promoting innovation';
- 'Rebalancing the distribution of functions in the metropolitan area';
- 'Re-organizing mobility and accessibility'; and
- 'Improving urban quality as a resource for development'. (Firenze 2010 2001:23, translation by the authors)

The very wide-ranging objectives are supported by massive urban marketing and promotional multimedia campaigns that are covering up a patchwork of projects already planned or in the course of realization before the drafting of the Strategic Plan itself.

The Structural Plan is based on the regional law n.1/2005 that pledges sustainable urban development and public participation, defining strategies for land use and development over a long period of time. The Florence Structural Plan (see Comune di Firenze 2007a, 2007b), though yet to receive final official approval, assumes three definitions of Florence on which projects are based. First is the 'brand name'. Here, the city is a 'modern myth', the name of which, 'best known in the world, recalls memory of beauty, elegance and good taste'; a place where 'historical and cultural heritage match well without any conflicts with contemporary daily life, [...] without turning the city into a museum' (Comune di Firenze 2007b:9, translation by the authors). Second is the 'global city'. Florence does not need to fight for a new role in the global market as 'its *missions* [sic] already granted the city a vantage point which will never be endangered' (ibid.:8–9, translation by the authors). The word 'mission' is used to emphasize vocational activities that in the course of time demonstrated a high degree of excellence in trade, arts and culture. Third is the 'city of good governance'. Florence is an open place for people and nations to meet and a key place for education, research and creativity as the 'invisible fabric of experimental initiatives' and innovation (Comune di Firenze 2007b:30, translation by the authors).

Beside the rhetoric of both plans, Florence's everyday reality is very different. Its public space, for instance, suffers dramatically for the branding of the city:

overwhelmingly affected by mass tourism, it undergoes a process of museumization and disneyfication. The position of Florence in the global market is not per se of good quality as it does not protect citizens from the deterioration of its living conditions. The creativity of Florence and its especially innovative subcultural productions struggle to survive and are not at all recognized by official cultural institutions: they are rather almost neglected, if not manifestly ostracized, and nearly disappearing (Paba 2001). The concept of Florence as the city of good governance (Comune di Firenze 2007b), a city that supports public participatory processes and eulogizes itself as pluralistic, clashes dramatically with the reality of an administration unable to put into practice consensual and 'non-violent conflict management' (Friedmann 2000:470).

Despite the potential elements of innovation contained in the new regional law, and the claim for a participative involvement of citizens, the actual outcomes of the planning process seem to go in a different direction, being strongly informed by consolidated power relationships and market pressures. What is often missing is the logical consequence between the objectives enunciated by the plans and the actual projects put in place.

For instance, how would such policies and projects 'protect and reinforce the identity of the historical centre and the city as place for residence and high quality handcraft', 'to revitalize the city as a centre of cultural production, formation and of technological innovation', 'to improve environmental quality and the mobility' (Firenze 2010 2001:34, translation by the authors), to quote just some of the Strategic Plans' purposes. All these goals, as a matter of fact, the city is dramatically failing to achieve.

Tendencies transforming the city

Supported by the above-mentioned rhetorical discourses, urban strategies for Florence are redesigning the city through three main and connected tendencies: first, the decentralization of functions traditionally located in the centre, as university, law courts, military headquarters, administration and residence, freeing a huge amount of high-valued buildings and making them potentially available to the ruling economy of leisure and tourism. Second, the development of new polarities in the periphery. They absorb the last available spaces on the fringe where zoning had attracted land speculators and large real estate interests in the past. Third, the reorganization of mobility and transportation according to the decentralization logic, which prioritizes managing the flows of people and goods to and from the historical centre. This ongoing transformation demands an infrastructural network adjustment that has swallowed the largest amount of public investments in recent years. Its capacity to be leverage for urban regeneration and improvement of the quality of life raises public debate.

For years, Florence has been suffering mobility problems, and its province counts the highest number of private cars per inhabitant in Italy (Agenzia Regionale per la Protezione Ambientale della Toscana 2007). The implicit limits of the historical fabric, together with inadequate public transportation produce overwhelming car traffic, distressing the mobility into, out of, and around the city. The district inside the former city walls is the object of desire of consistent fluxes at the core of a radial metropolitan system, and the boulevards' ring around it is a substantial bottleneck, slowing down all traversal movements across the city. The whole city is also a critical bypass for the national fluxes of goods and people but is permanently at the edge of congestion and paralysis. Considerable regeneration plans for mobility issues propose a third lane for the motorway, the crossing of the city with a high-speed train and related new station, tramways, and parking lots around and inside the city centre. Most of these infrastructural solutions, defined as 'non-negotiable' (Comune di Firenze 2007b:43), will have a strong impact on the prestigious urban and natural landscape. These strategic projects have been handled by institutions with a top-down approach, cutting off any possible debate about alternative solutions. The public administration is privileging oversized technical solutions, in order to attract high investments in financial terms by favouring the interests of construction companies possessing a powerful voice in the decision-making process.

Public–private partnerships

Mobility and parking surveillance have become a significant employment sector in the economy of the city and a conspicuous source of income for the municipality and the enterprises connected to mobility issues by PPPs. The mobility plan for the city delivers a system of new underground parking lots all around the city centre, a strategy that deserves some criticism. The first critical point of this operation is to reinforce the excessive amount of private traffic, instead of desaturating the central location by granting priority to alternative forms of mobility. It is self-evident that increased hosting capacity of vehicles around the historical district is a contradictory strategy for limiting the already excessive traffic pressure and pollution.

A second critical point regards the procedure chosen to realize those plans and their results in terms both of efficiency and profits for the public finance. Infrastructural operations in this field are directed by the Municipality of Florence through two different yet interconnected organizations: Firenze Parcheggi and Firenze Mobilità. The first is a joint stock company whose main shareholder is the Municipality of Florence itself. It manages an increasing number of underground parking lots and the extensive system of surface pay-toll car parks in the city. In addition to the direct revenues of parking fees, it also benefits from 14 per cent of the revenues from parking fines. The second, Firenze Mobilità, is a holding company expressly created to respond to a call for developing new underground car parks. The main shareholder is Baldassini Tognozzi & Pontello (BTP), the biggest building enterprise active in Florence, together with Firenze Parcheggi itself, the Chamber of Commerce and other institutions. The project financing architecture foresees Firenze Parcheggi paying to Firenze Mobilità – for a certain amount of years – the rent of all the parking facilities, even of those actually underutilized.

If we espouse a liberal agenda to manage public functions through private enterprises, we at least expect effectiveness in producing revenues. Instead, Firenze Parcheggi is in constant budget loss. The main reason for the debt is due to money owed to Firenze Mobilità for the newly built parking facilities, the income capacity of which is lower than their costs. The situation is simply perverse as the debt is actually contracted with the banks that are at the same time shareholders of the company. As a matter of fact, the Municipality of Florence, sponsoring the creation of two private companies to *better* carry out the construction and management of a hypertrophic system for private mobility, spends a huge amount of public money in balancing the shortfalls of the PPP, guaranteeing the profit of the private stakeholders instead of benefits for its citizens. An emblematic example is the case of Fortezza da Basso, a masterpiece of renaissance architecture transformed into a fair ground, close to the main railway station. Already in an advanced construction phase, public opinion acknowledged that part of the planned structure was going to emerge in front of the historical fortress of Giuliano da Sangallo. A campaign against the project was raised, forcing the administration to stop construction and to resize the project.

As a consequence, part of the newly built facility had to be pulled down due to its poor design. At the same time, the building company has not been able to excavate all the three underground floors, for the unexpected appearance of groundwater. The Municipality of Florence is now paying the compensation for the vanished profits of Firenze Mobilità (10 million and 200 euro) (Ferrara 2008). Citizens, who already harshly criticized the project from the early stage for the impact in the historical area, now feel as if it is contributing to the burden by paying for the mistakes of the PPP through public taxes. Overall, the malcontent and criticism towards Firenze Parcheggi's initiatives is increasing in the city.

Recently, the Procura della Repubblica (National Prosecutor's Office) opened up a file to investigate the case of this Florentine PPP for the case of Fortezza da Basso and also for other public works, including among them the one of Piazza Ghiberti, in which civil servants of the municipality and representatives of Firenze Mobilità are accused of bribery, corruption and public fraud (Gomez 2008; Selvatici 2008a, 2008b).

The case of Piazza Ghiberti

In the Santa Croce neighbourhood, BTP won the tender to realize one of the biggest underground parking facilities owned by Firenze Parcheggi. The parking space is located beneath Piazza Ghiberti, in a strategic position between the Viali (the city centre boulevards' ring) and the historical inner city. The four-floor underground parking lot was successively reduced due to excavation difficulties. Despite this big change in the plan, Firenze Parcheggi paid Firenze Mobilità 10 million euro, exactly the sum initially agreed to pay four floors instead of the two actually realized (Selvatici 2008a). Most residents strongly criticized this project, which confirmed the public scepticism towards the PPP and the mistrust towards any Firenze Parcheggi plan.

In 2005, following the disgruntlement, the municipal administration launched a participatory planning workshop for a new Piazza Ghiberti (see Figure 6.2) in the framework of a larger participatory programme called 'Florence Together'. The

Figure 6.2 Piazza Ghiberti.

Photograph by Laura Colini.

workshop was dedicated to an international competition funded by the Firenze Parcheggi, which had just finished the underground parking facility underneath Piazza Ghiberti, and were about to complete the pedestrian surface with pavement and a lighting system. The workshop participants were asked to respond to a consultation for the design of a public square where the municipality proposed the relocation of the antiquity market already existing in the neighbourhood. Both the competition and the workshop were supported by the municipality.

The participation process seemed to be biased from the beginning. The project for the surface of the Piazza is financed by the Firenze Parcheggi, which is eager to regain public support in the city. The local authority is seeking to recreate a dialogue beyond the rancorous attitude of its citizens, but yet proposing a predefined solution, which may jeopardize the honest spirit of a participatory decision-making process. Moreover, the workshop was not designed to contribute to the future development of the neighbourhood as, officially, it was intended to focus exclusively on the design competition of Piazza Ghiberti.

This piazza is probably the largest public space in the historical area and an important core of Santa Croce neighbourhood. Santa Croce is known to be one of the traditional social hearts of Florence (Pratolini 1943) which today counts on an active and ethnically diverse resident population, a local neighbourhood administration, socio-cultural associations, a number of activists organizations, and independent groups. Beside the historic residential houses and small shops, the area hosts a mix of vital functions such as *cinema d'essai*, religious centres (synagogue, mosque and more than one catholic church), the University of Florence (Faculty of Architecture), a recent social housing programme in Piazza Madonna della Neve, the grocery market of Sant'Ambrogio, the well-known antiquity market dei Ciompi and education and social facilities. Some changes in the life of the area such as the large presence of students, the slowly disappearing arts and crafts laboratories and shops, the vacating of the local newspaper building La Nazione, bought by a supermarket chain and the relocation of the court house to the newly built Palazzo di Giustizia in the outskirts, opened up a debate about the future scenario of the neighbourhood. Discontent affects both the inhabitants as well as the working population of this area, who demand a voice in the future development of the neighbourhood. Long-time tenants tend to move out of the area due to the high costs of dwelling there. Crystallization of the city centre for mainstream purchasing activities makes it hard for arts and crafts activities to survive. Moreover, vendors of dei Ciompi market declare that the number of customers has been dramatically reduced after the creation of a pedestrian area, which is not supported by sufficient and efficient public transportation.

Surrounded by the Faculty of Architecture, the street that separates it from the grocery market, La Nazione building, and some residential buildings, the Piazza became a large empty space after the demolition of some shanty houses for the construction of car parking. Now, the Piazza Ghiberti is only an empty space in the tight-knit texture of the historical centre with a great potential for neighbourhood life.

The vacated La Nazione building will be reused some time in the future and its new functions will surely be connected to the square. The Faculty of Architecture – constantly under enlargement – can benefit from this large outer space. The grocery market could profit from an extension or connection with trading activities and the residents could benefit from a rare open-air recreational facility in the centre. In addition to seasonal markets and fairs, the Piazza also welcomes different activities and proposals such as those carried out by a group of squatters who occupied some of the empty residential buildings facing the Piazza, and the local radio willing to set up a radio station in the Piazza as part of a collaboration with the university. These and other possible functions are by no means detached from the life of the whole neighbourhood.

The launching of a participatory process offered the opportunity for discussing and debating some of these issues, in order to create a shared vision for the future of the Piazza, bringing together all the stakeholders and making visible long-time conflicts, moving toward a common resolution. A precondition for the workshop was an agreement between the municipality, the residents joining the first meeting, and the non-profit foundation Fondazione Michelucci, in charge of facilitating the workshop. The condition was that the workshop was solely to discuss strategic guidelines for the whole neighbourhood development, including the new Piazza Ghiberti, the design of which will be realized through the public competition.

Despite the true commitment of the citizens and the Fondazione Michelucci in proposing both guidelines for neighbourhood regeneration and the Piazza Ghiberti, the obligation of the local administration towards the citizens remained closely limited in the frame of the design competition, and in the agreements previously accorded to Firenze Parcheggi. Since the conclusion of the workshop and the successful accomplishment of the design competition at the end of 2005, there has been no sign of launching the regeneration plans or projects, either in the neighbourhood, or in the Piazza Ghiberti.

Moreover, at some point in the workshop, the municipality revealed key information that was not transparent from the beginning, a hitch supposedly due to a miscommunication among municipal departments, which risked the trustful relationship that had been patiently built among the workshop's participants. As a result, the municipality seemed to be caught between a willingness to regain the trust and dialogue with its citizens, the clogs of the institutional machine and the pressure from mostly private investors; they were unable to balance them for the sake of the citizens' welfare.

Public partial interventions such as these may risk reaffirming the division of interests against a common vision for the city, hiding decisions driven by exclusive monetary interests, or delaying ad infinitum and finally moving to other decisions pending key questions.

Conclusion

A comprehensive portrait of the urban renaissance of Florence is a controversial matter. On one hand, the city is clearly successful in confirming the rank that it has reached in a global landscape of cultural heritage capitals and in perpetuating its traditional attractiveness. Paradoxically, this success has a catastrophic effect on the social and cultural life of its inhabitants, on the everyday life conditions, and on the capacity to renovate the conditions of creativity that made Florence able to become that extraordinary cradle of beauty and art in the first place. Instead, the renaissance of real estate developers, fashion traders, tourist operators and other privileged managers of commodified spaces and facilities has led to the dispersal of residents, students, artists, craftspeople and intellectuals and to the debacle of contemporary culture. Today the city still demonstrates its ability to maintain its heritage in an authentic way and distribute with dignity its welfare among old and new citizens, providing a good standard of education, health care and social services. Yet the lively, proactive relationship envisaged between citizens and the city, as the precondition for a real urban renaissance in Florence, is far from being realized.

Acknowledgement: The authors would like to say thanks to Mark Kammerbauer.

7 Winners and losers from urban growth in South East England

Bob Colenutt

Growth is the holy grail of UK government urban and economic policy. The South East of England in and around London is bursting at the seams. The government view is that in order to sustain this engine of the UK economy and maintain national economic competitiveness, a step change in housing supply is needed (Barker 2004:1). To make this politically acceptable, urban and regional planning policy has been placed within a framework of 'sustainable communities'.

The sustainable communities programme is long term and is a comparatively recent policy position (Office of the Deputy Prime Minister 2003). Yet the underlying principles are well established, housing targets are confirmed, plans for several large-scale housing developments are at the first stages of implementation and new subregional delivery agencies are in place.

At the local and regional level, authorities are equally enthusiastic for new housing and business. Despite environmental concerns and fears that there may not be sufficient funding for statutory services, most are active partners with government in planning for housing growth, seeing it as an opportunity for prosperity and image change.

Although there is an intensive public and academic debate about the merits of growth versus no growth, and about widely discussed urban form issues (Town and Country Planning Association 2007), there is little attention paid to the costs and benefits of the growth plans. Who are the winners and losers?

The area chosen to examine the distributive effects of the growth agenda is Northamptonshire, part of the Milton Keynes South Midlands (MKSM) growth area (see Figure 7.1). Northamptonshire is a comparatively rural county in the middle of England, 60 miles north of London, straddling the M1 motorway. It has a population of 600,000 concentrated in a network of former industrial towns now circled by housing estates, business parks, retail centres and road systems. The largest town is Northampton with a population of 200,000. Famous for its boot and shoe industry in the early twentieth century, the town has undergone an economic and population transformation since the end of the Second World War (Greenall 2000:29).

Growth policy for the South East

There has been housing growth outside London for decades driven by differing strategic planning and housing policies. Central government intervention in South East planning in the post-war years was motivated by planned population

Figure 7.1 Map of MKSM growth area.

Source: Government Offices for the South East, East Midlands, East of England 2005.

overspill from London. It included building a circle of new and expanded towns around London from the 1940s up to the 1960s. The largest of these new towns, Milton Keynes, was started in 1967 and continues to expand at a rapid rate today. Although much of the housing stock in the new and expanded towns has now been sold, most of the housing units were built for rent to meet the needs of those in run-down housing in London and Birmingham.

The policy behind the current growth plans is almost the reverse. It is led by demand for housing for sale. To facilitate the supply of new private housing, central government is playing a key role as plan maker and funder. It has provided the strategic framework for growth through regional policy guidance; it is assembling brownfield land for development using its own national land and housing agency; and it is funding transport infrastructure (mainly roads) to enable growth to take place. Government concern about market housing supply goes back some ten years, stemming from rapidly rising house prices in London and the South East. Political pressure has mounted from first-time buyers priced out of the market, and from house builders and landowners claiming that not enough land is being released for development. These concerns prompted an investigation directed by the Treasury in 2004 of the housing market by an economist, Kate Barker. Her report concluded that only by increasing housing supply very significantly could house prices be restrained and demand be met. Barker suggested that supply was restricted by lack of land for new housing, and planning delays by local authorities. She concluded that the planning system should be speeded up to enable developers to obtain planning permission more quickly, and central government should intervene to make more brownfield land available, thus helping to bring down house prices.

The Sustainable Communities Plan (SCP) launched by the Blair Government in 2003 under the direction of Deputy Prime Minister John Prescott became an instrument to deliver the Treasury plan. The aim of the SCP was to designate four 'growth areas' around London (see Figure 7.2):

1   Ashford in Kent to the south of London;
2   London-Stansted-Cambridge-Peterborough corridor to the northeast;
3   MKSM (including Northamptonshire) to the north; and
4   Thames Gateway to the east of London.

Most of the growth would be in the form of 'urban extensions' around existing towns rather than in new towns or in entirely new settlements. In other words, they would be attached to existing towns in a planned way, to prevent building in the countryside.

The key message was 'sustainable communities'. New housing would not be simply mass-produced, single-tenure housing estates with limited services, but mixed communities that would be socially, economically and environmentally sustainable. More recently, there have been proposals to go further to meet concerns about climate change by building up to 10 zero-carbon 'eco-towns', many in the growth areas. Government also said that by 2016 all new homes will have to meet zero-carbon standards.

In the four South East growth areas, 200,000 additional housing units above current levels of growth are planned by 2016. They are accompanied by major public investment in roads, schools, affordable housing and other infrastructure. Alongside this, there will be significant private sector investment in industrial and commercial buildings (Government Office for the South East, East Midlands, East of England 2005).

Shortages of affordable housing (with the definition extended to include housing for first-time buyers, not just those on housing waiting lists) are recognized.

Actions to Achieve and Manintain Sustainability

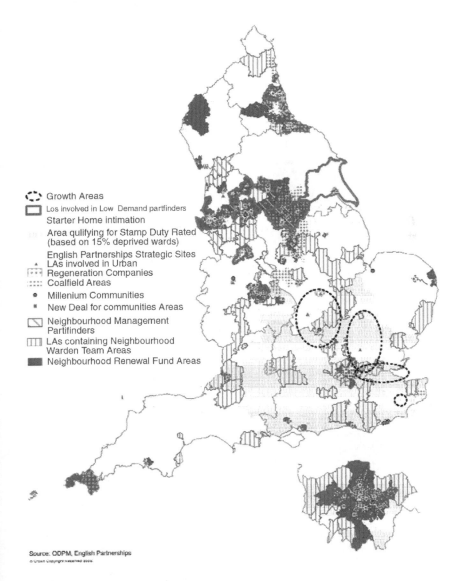

⊂⊃ Growth Areas

☐ Los involved in Low Demand partfinders
Starter Home intimation

Area qulifying for Stamp Duty Rated
(based on 15% deprived wards)

English Partnerships Strategic Sites
▲ LAs involved in Urban
Regeneration Companies
⁞⁞⁞⁞ Coalfield Areas

● Millenium Communities

✳ New Deal for communities Areas

◨ Neighbourhood Management
Partifinders

▥ LAs containing Neighbourhood
Warden Team Areas

◼ Neighbourhood Renewal Fund Areas

Source: ODPM, English Partnerships

Figure 7.2 Map of growth areas in South East England.

Source: Office of the Deputy Prime Minister 2003.

The Barker report acknowledges that 'provision of social housing has not kept up with need; the number of newly built social houses for rent has fallen' (Barker 2004:89). But funding for local authorities and housing associations to build the large number of new social homes required has not been forthcoming. Government is relying on planning gain to leverage the remaining funds for social housing. Typically, local authorities require around 20 per cent affordable housing on large sites and seek 'developer contributions' (Section 106 payments) for this and other infrastructure costs. But research by the Joseph Rowntree Foundation found that of housing units built between 1997 and 2000, only about 10 per cent of all new units built were 'affordable units' funded from planning gain (Crook *et al.* 2002).

The property lobby

As planning restrictions are lifted and land is released for new housing, landowners, builders, developers and estate agents have been significant beneficiaries of the growth agenda. Landowners have made large profits from increases in land value arising from land being granted permission for housing.

So large are profits to be made by house builders and landowners that the industry has so far been willing to accept the payment of a 'tariff' for each new house built in some of the growth areas. The tariff (which varies from area to area) is a payment under Section 106 of the Planning Acts, assessed on the number of houses to be included in a development, then pooled by the planning authority to contribute to infrastructure costs. The tariff ranges from £18,000 per house in Milton Keynes to £10,000 per house in other areas.

Even with the tariff, the risks to the building industry are minimal. Developers allow for a 20 per cent profit on any development, and very few undertake development at less than that. House builders and developers are very well-organized lobbies, pressing government continually for concessions on planning, capital allowances, and financial incentives to build on brownfield land. Their most strident demand is for a 'reduction in the stranglehold of bureaucracy' (British Property Federation 2004).

However, government emphasis on building first on brownfield land is not popular with these lobbies, as this brings potentially higher costs of land preparation on sites in less attractive locations. But the property industry has already managed to gain important concessions to allow building on greenfield land with government now considering relaxing the building restrictions within parts of the important amenity area around London called the 'Green Belt'.

These examples show how far the politics of property development have changed since the 1980s. Developers and house builders are strategic partners at every level represented on government advisory committees and local authority development companies. An interesting recent case demonstrates the power of the house builders. Some local authorities under what is known as the 'Merton Rule' have negotiated a far-sighted 10 per cent 'energy from renewables' condition on all new housing units. This requires builders to reduce carbon emissions by 10 per cent through the use of renewable energy sources. But the house builders have successfully lobbied to abolish the Merton Rule. Instead they want a 'national strategy' to enable the phasing in of zero-carbon policy in 2016, thus deferring action on these issues for many years – if ever (Seager 2007).

Why growth in Northamptonshire?

Northamptonshire is typical of counties to the north of London because it has experienced successive waves of planned expansion since the 1950s. For example, Northampton, Kettering, Wellingborough and Daventry were designated expanded towns and overspill towns, while Corby was one of the first post-war new towns.

The legacy of this period is a predominantly low-skills area, in spite of relatively full employment and steady growth. The current profile of light engineering, transport and distribution, food and drink manufacture, and back-office financial services has perpetuated a relatively low-wage economy. Thus, educational achievement and progression to higher education are below the national average (Kerrigan 2006). Town centres are in need of fresh investment and it has been difficult to attract high-value businesses. The central issue is that parts of the county, particularly in the north around Corby are 'stuck' in a low-wage, low-skills cycle.

From a pure land development point of view, Northamptonshire has a lot to offer. There is plentiful land at reasonable prices, compared with counties closer to London. There are very good road and rail connections to London, and many towns, particularly Northampton itself, want an improvement to their image – and extra funds to pay for services. Growth is an opportunity to compete for investment and bring in residents with more spending power (North Northants Joint Planning Unit 2007). The key question is whether growth is, at the same time, an opportunity to tackle the underlying social and economic problems of the county – this perhaps is the real test of the SCP.

The politics of growth in Northamptonshire

When the SCP was announced, many of the local authorities were Conservative controlled with a strong 'protect the countryside' agenda. The Labour Government in London with its plans 'to build over the countryside' was seen as the enemy.

A STOP campaign based in rural areas was organized with several local authorities showing sympathy. The campaign was opposed to any encroachment on the villages in the county. It was allied to the Campaign for the Preservation of Rural England and Friends of the Earth who believed that the growth agenda would lead to building on greenfield land and would therefore be unsustainable environmentally. This response produced complex local government politics. With one exception, the Conservative-controlled districts in the south and east of the county wanted to stay out of the growth plans, while the larger urban centres of Northampton and Corby, and the County Council itself, all Labour controlled at that time, actively sought inclusion.

The County Council, with a less parochial vision than the districts, acted as broker. It convinced other local authorities to see the growth plans as a chance to attract government funding for services and run-down town centres. A more difficult task was getting agreement among local councils to have a Development

Corporation in the west of the county including Northampton to break through the conflicts between local authorities, by taking over their local planning control powers. There was some anger about this, and a quite reasonable scepticism that government would not come up with the money for public services and that the result would be new housing estates with no services (and higher costs to local authority rate payers).

The way around this for the proponents of growth was to say that local council members would be on the board of the Development Corporation. Also, if the local authorities were working with regional agencies and government, they could ensure that funding for infrastructure and services went hand in hand with housing growth. Some local councils believed they could 'manage' growth and attract funding from government that would regenerate their communities.

The case for and against a West Northants Development Corporation was argued out at a special Parliamentary Committee in 2005/2006 who found in favour of the scheme. The government's case was that the Development Corporation would not be like those of the 1980s, because it would have a strong commitment to local consultation. In the north of the county, local authorities (including Corby) opted for a joint partnership and planning board (the North Northants Development Company) to retain local democratic control.

The government prepared a plan for the MKSM area covering Milton Keynes, Bedford, Luton and Northamptonshire (Government Office for the South East, East Midlands, East of England 2005). The targets for Northamptonshire were 90,000 new houses by 2020 (the largest share of growth in the MKSM area); 80,000 new jobs; dozens of new schools, new roads, parks and open spaces; the promise of town centre revitalization; and additional affordable housing (see Figure 7.1). Growth for Northamptonshire was described as a 'step change' in aspirations and investment.

Priors Hall urban extension

Housing growth would take place principally in urban extensions. The extensions would have between 3,000 and 5,000 houses complete with a range of services, schools, parks and shops. Typically, the estates would have mixed tenure (generally 20 per cent 'affordable' and social housing), around 20 per cent flats, 60 per cent low- to middle-income private houses, and around 20 per cent 'executive' housing. They were intended to be exemplary new settlements led by best practice in 'design coding' and sustainable development principles.

Priors Hall on the edge of Corby is typical of these urban extensions. It is worth looking at in detail to understand its purpose – and its likely impact on the neighbouring town of Corby. Covering 400 hectares, Priors Hall is a £55 million scheme of 5,100 homes plus schools and other services. It received outline planning permission in April 2005. The land is owned by a single owner, BeeBee Developments. Most of the housing will be for sale, with only 10 per cent 'affordable' (shared ownership and sheltered housing). Some of the Section 106 (planning gain) funds from the developers will go into education and community facilities in and around the site. The neighbourhood designs and housing styles are attractive, showing concern for quality and the environment. It aims to be environmentally sustainable.

Corby itself is a former new town with a steel works that closed in the 1980s with the loss of 6,000 jobs. It has reinvented itself over the past 15 years as a successful distribution and food processing centre, although it has the lowest average wage and skill levels in the county. Many Corby residents live on unpopular former new town housing estates, creating a demand for better quality social housing and lower priced private housing. The new economy has attracted in-migrants particularly from Eastern Europe. Predominantly workers in low-skill industries, they too want access to affordable housing.

The landowner, the local council and the North Northants Development Company are seeking to change the fortunes of the town by building an infrastructure of business and schools that will attract higher income purchasers for the new housing in Priors Hall, raise land values, and stimulate a cycle of improvement that will break the low-wage, low-skills cycle. The question is, will this strategy succeed?

A recent housing market survey for Corby itself up to 2021, showed there was demand for 78 per cent owner-occupied housing (at the lower end of the market) and 20 per cent affordable housing (BeLa Partnership 2007). Yet in fact the plans for Priors Hall show that, to ensure levels of viability demanded by the landowner and developer, early phases of new housing will be targeted at middle-income and executive markets, with most of the new homeowners expected to come from outside Corby, attracted by better value for money for housing than in suburban London itself (Cowans *et al.* 2007). A railway station with services to London is to be reopened which will encourage commuters to London and thus greatly assist this process.

There will be indirect benefits from improved facilities in the town funded from Section 106 agreements such as new school buildings, social housing renewal, and a construction training centre. It is also hoped that residents of Priors Hall will shop in the town centre and thus uplift the retail offer there. The owner of the retail centre in the town has responded by planning a new shopping centre. Also, residents of Corby are getting assistance from other urban regeneration programmes for the demolition and rebuilding of some of the badly designed housing estates of the new town, alongside a small programme of neighbourhood renewal in the deprived Kingswood area.

Yet Cowans, Robinson and Meikle (2007), who advocate the strategy of attracting higher income home buyers, warn that existing programmes will not be enough to ensure the benefits of Corby's renaissance are widely spread, and that direct channelling of funding into Corby will be needed to avoid the creation of a 'doughnut' effect – that is, where low-wage households are clustered in the town centre, and the better off are in the suburbs. Where this funding will come from is uncertain. Not much can be expected from the development tariff outlined earlier, since most of this money will be absorbed by strategic infrastructure, particularly road building and schools, to enable the urban extensions to take place.

More importantly, while growth will deliver new house building, less thought has been given to the investment in education and skills that must accompany economic development (Colenutt *et al.* 2007). The planner's concept of economic sustainability is simplistic – to keep housing numbers and jobs in balance, and to target 'knowledge economy' inward investment. This rather avoids the challenges of low incomes and skill levels, out-commuting and the massive investment in education and skills and employer engagement required to create a sustainable Corby community. The planning documents and regeneration strategies have little specific to say about how to break the low-wage cycle in the town, or how a knowledge-based economy can become a reality. The strategy relies more on the hope of a trickle down of economic prosperity (Cowans *et al.* 2007:35). Moreover, environmental sustainability is at risk as new residents in the first few phases of Priors Hall commute out by car or rail to towns and cities with higher value employment.

The irony is that this important debate may well not be heard at all in the Department of Communities and Local Government or in the Treasury back in London. Urban extensions like Priors Hall will be judged not by the contribution that they make to transport, jobs or to sustainable living, but by the numbers of housing starts and completions. Housing supply will go up – but the question of who is excluded from the renaissance will go unheeded.

Conclusion

It is hard to find evidence from the example of Corby in Northamptonshire that the SCP for the South East will create communities that are socially, economically or environmentally sustainable.

Housing growth in the region has been packaged as sustainable development but, in fact, this obscures the reality of a lack of integrated planning of the growth towns. In Northamptonshire, the new housing estates do not spring primarily from local (or subregional) housing need but wider demand for middle-income housing in the South East. Economic planning for these towns is taking second place to maximizing housing supply. There is an all-but-open admission that many of the new homeowners will commute long distances to their work in London or elsewhere.

How far this analysis can be extended across the four growth areas is unclear, although there are similar experiences elsewhere. In Cambridge, for example, where there is significant demand for higher income housing on the back of a high-value local economy, there are concerns that housing and jobs are getting out of balance as new housing is being planned to meet demand across the South East, rather than demand generated by Cambridge itself.

What is interesting is that the issue of who benefits from the growth agenda has barely surfaced in the mainstream politics of growth. This debate has focused mainly on the rights and wrongs of growth, or of building on greenfield sites. This includes a number of vigorous local campaigns against proposed 'eco-towns'. On one side is an uneasy coalition of anti-growth groups and conservative 'Not In My Back Yard' authorities; on the other, Labour ministers argue that there is a need for more housing and that it has to go somewhere (Cohen 2007). Concerns for equity have been almost completely overlooked.

The property industry is the major winner. It has the backing of both local and central government, is able to obtain the planning consents it wants, and is using its considerable influence to alter housing and planning policy at central and regional government level in its interests. But those without a property stake, on low wages, in social housing and unable to afford a mortgage are the clear losers.

The alternative is not to oppose growth, but to challenge the government notion of 'sustainable communities'. If the planning of regions and urban areas gave social, economic and community need the same priority as the race to follow market demand, the outcome would be quite different. In this context, there could be a more honest debate about what sustainable communities could mean – and how to achieve them.

Part II

On local limits to regeneration strategies

When fish sing in Brussels

Images and text by Ruth Pringle

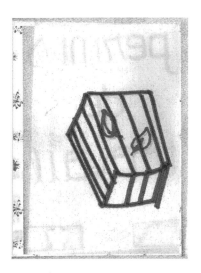

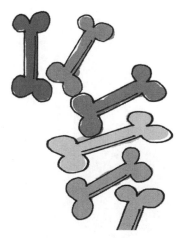

Number 102, Rue de Laeken is an exotic fish and pet supply store. It lies within a poor, largely Congolese neighbourhood adjacent to Brussels' city centre. The facade of the pet shop is covered by thick, cracked paint. Not just peeling, you feel like by throwing a stone you could start an avalanche. A large plastic orange fish hangs by a fin from the third floor balcony; its one eye stares down the street.

In the left-hand window, food bowls and retractable leads rest upon Astroturf, sheltered under a sparse plastic jungle and fading photographs of happy pets. The right-hand window presents a rainbow of different coloured chewy bones. A variety of toys wait like beady-eyed votive figures for a slobbery, doggy fate. In the corner, a red plastic birdcage keeps captive two yellowing plastic leaves.

Next door is a contemporary art gallery. Once an empty shop, it has now been refurbished in a sophisticated monochrome of glass, chrome and vinyl letters. In the window a thin (also monochrome) man on a ladder adjusts a spotlight and then sweeps his fringe from his forehead.

The exhibition opens that evening. Tiny, exotic fish with rippling dorsal fins and long noses have been imported directly from Africa to be artfully arranged inside immaculate glass tanks. Each tank has its own underwater microphones, and the electrical charges that these remarkable fish produce are being transmitted through loud speakers as a final sound check. Later, these little fish (the source of the synthetic, clicking noises being magnified by powerful speakers) will unwittingly star in a live electronic concert – aided by a mixing desk, sampler, the latest Apple power-book computer technology and two experimental musicians from Germany.

Next door in the pet shop, the fish swim restlessly, disturbed by the vibrations and a chunk of weathered paint drops to the ground.

Leaves: Original sketch by Ruth Pringle. *Bones*: Original sketch by Ruth Pringle. *Singing fish*: Photograph by Ruth Pringle.

8 Renaissance through demolition in Leipzig

Matthias Bernt

When commentators discuss the 'urban renaissance' of Leipzig, which is the regional capital of Western Saxony and a city of roughly half a million inhabitants, opinions are often split. The reason is that the city, with a majority of the housing stock built prior to the First World War, is, on one hand, a showcase example of a 'European city' with historical buildings, picturesque squares and galleries, lots of historic identity and urban atmosphere. Consequently, Leipzig is often seen as a 'comeback city' and one of the few East German cities that made it. However, a closer look removes many illusions. Numerous places outside of the city centre proper look like they have survived a disaster. Elegant abandoned buildings, empty office complexes, and blighted industrial brownfields on a frayed perimeter are the irrefutable signs of profound urban crisis.

The aim of this chapter is not to decide which of these interpretations is correct. Rather, it takes the contradictory situation as a starting point to discuss limits to urban renaissance. It considers why regeneration initiatives in Leipzig have not yet fully succeeded and shows which problems occur when 'standard' models of regeneration are applied in a situation of low demand and population decline.

Leipzig as a shrinking city

When discussing regeneration politics in Leipzig, it is extremely important to understand the specific background of urban development – which is one of long-term decline. As early as the 1930s, when the population of Leipzig reached its peak at 700,000 and the city was one of the main industrial, scientific, commercial and cultural centres of the German 'Reich', the city experienced population losses. Since then, a continuous downturn has characterized the city's development. When Germany was reunited, this recession trend was accelerated for another decade. Over a span of ten years, Leipzig lost about 80 per cent of its industrial jobs. In an unprecedented process of plant closure and mass lay-offs, about 90,000 jobs were lost in just three years, between 1990 and 1993. The development of the tertiary sector did not compensate these losses, and unemployment rose to about 20 per cent, which has stayed constant ever since. Paradoxically, the economic recession was coupled with an immense boom in building. Altogether, in the 1990s about 34,000 new homes were built in Leipzig

Table 8.1 Housing types and vacancies in Leipzig, 2002 and 2005

Date of construction	Housing stock		Vacancies 2002		Vacancies 2005	
	2002	2005	Total	%	Total	%
Pre 1918	111,598	110,626	27,000	24	20,000	18
1919–1948	58,042	57,625	11,000	19	11,000	19
1949–1990	100,228	97,887	15,000	15	12,500	13
After 1991	46,535	49,889	2,000	4	1,500	3
Total	316,763	316,027	55,000	17	45,000	14

Source: Stadt Leipzig 2006.

and its surroundings, mostly single- and multi-family homes in suburban areas. As a result, the city emptied out; in ten years, the core city lost almost 100,000 inhabitants, one-fifth of its whole population.

Consequently, the balance of Leipzig's urban structure came under immense stress. With renting forming the dominant form of tenure, the oversupply in the housing market stimulated many residents to move to better apartments, resulting in a tremendous increase in vacancy levels, even in well-located and good-quality houses. The most visible feature of this development was the high number of vacant apartments (see Table 8.1). There were a total of close to 60,000 empty homes in Leipzig at the turn of the millennium, about a sixth of the total. It was clear to everyone that Leipzig was a city 'whose dress became too big' (Kabisch 2001).

The consequence was, however, not a 'doughnut effect' as was typical for many American cities. As vacancies were widespread rather than concentrated, Leipzig's urban form was instead being perforated. Alongside beautiful renovated buildings, there are buildings reduced to a state of ruin, with *Baulücken*, vacant lots on which ramshackle buildings stood that have been recently demolished or have collapsed on their own. This makes the urban face one in which the old stands alongside the new, and the unbuilt lots border compact building blocks. The unified perspective of the historic Wilhelminian architectural period has frayed.

Taking the challenge: Leipzig's regeneration agenda

How are urban planners dealing with this state of affairs? Interestingly, the whole issue was entirely taboo (throughout all of Germany) until about the turn of the millennium. It was only in 1998 that Leipzig's city planners started to realize that many of the existing buildings could not be preserved. The discussions at this point in time circulated around the 'last quarter' of unrefurbished buildings without a realistic prospect for renovation. Based on these dramatic estimations, planners developed a threefold strategy. First, through finance from a new federal government programme, about 10,000–15,000 homes were to be demolished in order to cut down the excess in supply, restore market equilibrium and regain the confidence of banks and investors that Leipzig's real estate market would recover. Second, buildings to be demolished were to be chosen in such a way that value would be added to the affected areas. 'More green by fewer houses' was a famous slogan that summed up this objective. Demolitions in the periphery and a targeted thinning out were seen as key to achieve suburban qualities in the densely built Wilhelminian areas. Third, Selbstnutzerprogramm, a new 'owner occupied housing assistance programme', was introduced to capitalize on low prices in the property market, encourage home ownership and thus develop a new housing market in the inner city.

In dealing with decline, Leipzig thus developed a policy that supplemented traditional upgrading by adding measures aimed at a controlled reduction of the existing built environment. Although in practice this was very different from what is normally understood by 'urban renaissance' elsewhere, the policy framework drew heavily on the internationally accepted discourse of an urban renaissance agenda. As a consequence, urban development goals were not so much developed as a response to specific problems, as designed to capitalize on 'golden opportunities' towards urban regeneration. Typical of the way in which urban development was conceptualized by Leipzig's decision makers was a statement by the former head of the Building and Planning Department, Engelbert Lütke Daldrup. Defining the challenges and objectives in a publication entitled, 'Leipzig Plus/Minus', Lütke-Daldrup stated:

> The particular social asset that Leipzig possesses is of being a 'city of experiments'. ... The compact urban core of the city, the areas of perforation and the urban periphery are a varied theatre for innovative companies, creative individuals and their cultural blossoming. The cityscape of Leipzig is an unmistakable site of European culture, one that facilitates urbanism and identification. ... Leipzig will remain what it always was, a melting pot of old and new, of conversation and pioneering spirit, a sense of proportion and enthusiasm, in brief: an urban, European city.
>
> (Lütke-Daldrup and Doehler-Bhezadi 2004:120)

The interesting point here is the discursive shift which engaged globalized narratives of 'renaissance', 'urbanity' and 'creativity' to declare vacancies and demolitions an 'asset' which can help Leipzig successfully compete for 'innovative companies, creative individuals and their cultural blossoming'. Far from interpreting job losses, population decline, and physical decay as a problem for local residents and a responsibility for public authorities, the political focus gradually shifted towards an approach in which the city was seen as something of a playground for 'experiments' in which 'innovative' actors could try out new facets of urbanity.

Planned demolitions?

Translated into practical politics, most of these ideas hardly ever played out. Instead, a lag between the agenda of urbanity, creativity, and culture, on one hand, and the concentration of practical efforts on the simple physical reduction of the housing oversupply on the other, became evident. As a result, demolitions, and not an agenda of urban restructuring – however it might be defined – formed the central components of Leipzig's practical renewal politics.

The reason for this was mostly of an economic nature. For the affected property owners, vacancies of 15–25 per cent ate into capital resources, and led to dramatic losses of profit, losses of secure mortgages and ultimately, in the long term, to bankruptcy. Additionally, a lack of occupancy leads to a general devaluation of vacant sites, lower prices, a loss of mortgage value, and greater marketing expenditures for all those selling apartments. The crisis thus threatened not only the housing firms themselves, but also had a ripple effect adversely affecting suppliers, the construction industry and financial institutions as well. In the most severely affected neighbourhoods a downward spiralling was set in motion which led to a lack of maintenance, shop closures, a worsening of local supply and the development of so-called social problem areas. Low levels of occupancy were thus not only problems limited to one sector of a city but also the cause of a series of chain reactions affecting wide sectors in the urban fabric. Thus, the context under which regeneration was to be implemented was not one or several particular stigmatized neighbourhoods needing upgrades. Rather, the whole housing market was failing and needed repair in order for renaissance efforts to be successful.

It was within this context that the joint federal and state-level urban restructuring in the New Federal States programme was launched in 2001 – a first of its kind in the history of German housing politics – that provided subsidies for the demolition of vacant houses without substitution (see Bernt 2005). Similar to Leipzig's urban agenda, it embarked on a twofold urban development policy that simultaneously aimed to prop up the housing market and complete demolition of existing urban structures, so that the affected areas would be improved. However, after six years of practical application, it became more and more obvious that these goals had their limitations. The main obstacle was not the physical reduction of the housing stock, as by the end of 2007 about 200,000 housing units had been demolished. Rather, it was an inability to activate a broad upturn in inner city areas, and the widespread problems in the spatial coordination of demolition projects.

The main reason for these troubles was lack of cooperation amongst landlords. As it is up to the property owners where and whether plans for demolishing vacant properties can be implemented, urban restructuring on a broad scale is only possible when the owners cooperate. Although the housing industry as a whole has a strong interest in a balanced housing market, the tearing down of one's own property may not be particularly advantageous to individual owners. On the contrary, because (in theory, at least) the best situation for each individual owner occurs when all other property owners remove their inventory from the market until enough demand for the individual's own vacant buildings arises, there emerges a classic 'free rider' problem. This situation was summed up in a conference speech concerning an area in Leipzig's old city:

> Imagine that one finds a fine structure in an area of the old city. There are, for example, four blocks and one owner. This owner can remove several apartments or even tear down and replant an entire block. It's even possible to sell

this argument in economic terms by saying that in doing so I am stabilizing the rest of the existing buildings. The owner will support such measures. What happens, though, in such an area when one building belongs to Mr. Müller, the second to Mr. Meyer, and the third to Mr. Schulze? It's possible the ownership situation is not clear, perhaps there are things that should be torn down, but the old owner or the present owner has made claims to the property and has finally gotten his real estate back after a long struggle. Should he give up his property again and support the development of an open area …? Should one owner … sacrifice his own property that the value of the others grows? And his property is now a park that he still has to pay taxes and street-cleaning fees for?

(Beck 2002:82)

Realistically, it is misleading to suppose that all landlords plagued by low demand would be in the same boat. On the contrary, interests and capacities are very different from landlord to landlord and in this context, 'shrinkage' can only be implemented where the owner has an interest in the demolition of the property. In practice, this leads to a pattern of situations where some landlords willing to cooperate in demolition efforts sat cheek by jowl with others refusing to do so. Spatially, the conditions needed to implement urban restructuring in the New Federal States are therefore fragmented, not only on a city-wide level but also on a very micro scale. As a result, willingness to participate in demolition efforts is seen mainly from municipal and cooperative owners of large prefabricated concrete tower blocks and with small private owners of historic buildings that are so badly maintained that they were on the verge of collapse anyway. Whereas the former have enough 'mass' to cope with a loss of assets, and can profit from a release of 'existing debts' (resulting from loans granted to state-owned housing during German Democratic Republic times) for those apartments which are torn down, the latter can avoid costs for securing a building for which no tenants can be found.

In both cases, the spatial pattern of these interests is randomly distributed. While, for example, one owner is interested in receiving subsidies for demolishing a property, their neighbour to the right might reject this and renovate instead, while the neighbour to the left see-saws between taking subsidies or selling the house. The result of this uneven division of interests and strategies is a random distribution of possible demolition. As this condition determines the feasibility of implementing the urban restructuring strategy, the spatial fragmentation of conditions and interests led to an asymmetrical pattern of demolitions that has little to do with following a spatial masterplan. Contrary to the original intentions, urban restructuring thus arises as a random process that derives from a combination of differing types of ownership, credit worthiness and mortgage situations. Furthermore, compared with the overall development of Leipzig's housing market, it is indeed arguable whether or not demolitions for the sake of regaining housing market equilibrium can be more than a drop in the ocean. Thus, of a total housing stock of about 316,000 apartments, supported by 24.6 million euros of public subsidies, only 6,430 apartments have been torn down since 2001 (see Table 8.2). At the same time, 6,800 new

Table 8.2 Vacancies, demolitions, new constructions in Leipzig, 2001–2005

	2001	*2002*	*2003*	*2004*	*2005*
Reduction of housing stock	798	1,687	1,731	1,128	1,231
New constructions, including renovations	2,525	984	1,298	1,112	881
Total vacancies	—	55,000	51,000	50,000	45,000

Source: Stadt Leipzig 2006.

apartments were newly built or refurbished. The effect of demolitions on the over-all balance between supply and demand was thus minimal, and the vacancy rate was only reduced by a meagre 3 per cent (Stadtforum 2006).

In contrast, demolitions have affected most neighbourhoods, in many cases destroying historic urban fabric and damaging the urban form. Despite all rhetoric hailing the historic qualities and 'European' urbanity of Leipzig, numerous historic buildings have thus been demolished in the last couple of years. A non-governmental initiative has even listed the number of more than 30 listed buildings that have been torn down with public subsidies in the past years (Stadtforum 2006). Far from being embedded into existing urban structures and reacting to the difficult situation with local landlords outlined above, demolitions have often taken place in an ad hoc way and could thus add value to the affected areas only to a minor degree.

More green by fewer houses

Exemplarily, the problems in delivering urban renewal can be shown in the sector of green space management. Whereas the City of Leipzig has indeed used demolitions to gain more green space and improve urban environments, it is also in this field that fragmented ownership situations have considerably reduced the chances of successfully implementing its plans.

Nearly paradigmatically, this can be shown in Leipzig's eastern neighbourhood, a densely built Wilhelminian quarter which is burdened by a lack of green spaces, high traffic congestion, above average population losses and a negative image, the latter of which has been high on the renaissance agenda of Leipzig's planners. Consequently, when shrinkage became a new problem, Konzeptioneller Stadtteilplan Leipzig-Ost, a new 'Concept for the Development of Leipzig-East', was developed in 2001. One of the centrepieces of this plan became the Rietzschkebelt, an area along the Rietzschke (a small creek) in which demolitions were to be concentrated, mainly alongside existing green areas. Thus, existing free spaces could be connected and expanded to form a green belt cutting through the entire neighbourhood from the urban periphery to the immediate vicinity of the city centre. According to the plan, the new green belts were planned at the eastern edges of the neighbourhood at Wurzner Strasse. A park, known as Dunkler Wald (the dark forest), was to be created by tearing down a housing block, and then extend westwards, narrowing towards Rabet Park, which should also be expanded by demolishing bordering rows of houses. The intention was to create more free space with specific demolitions, thus making the neighbourhood more attractive.

Figure 8.1 Wurzner Strasse, Leipzig. 'Dark Forest' at Leipzig's East End.

Photograph by Matthias Bernt.

In practice, these plans did not work. It proved to be impossible to get all nec-essary housing owners to give up their properties, and it was virtually impossible to unite all of the lots needed to realize the plan. The demolitions and addition of green spaces have therefore come about in a rather random fashion instead of fol-lowing a set programme, and the plan is still at risk of remaining unfinished (see Figure 8.1). Moreover, to save money, the spaces that have been freed up through demolition have often become parking lots, dog-walking areas, or grassy fields. An increase in urban environmental quality is, therefore, rare.

Plans to achieve 'More green by fewer houses' have thus indeed led to an expansion of green spaces, but not in the intended way. Instead, the result is a hotchpotch in which renovated buildings and ruins, nicely developed pocket parks and garbage dumps, community gardens and dog runs all stand cheek by jowl.

Selbstnutzer.de: assistance for owner-occupied housing

A similar gap between ambitious plans and sluggish transformation is also char-acteristic of another area of Leipzig's urban renaissance policy: Selbstnutzer-programm, Leipzig's owner-occupied housing assistance programme introduced in 2001 with the aim to expand owner occupation.

The background here was the soaring suburbanization of the 1990s, which drew many Leipzigers out of town. As it is commonly believed among German urban planners that this high residential mobility was supported by the dominance of rentals in the core city in contrast to the higher share of ownership in the suburbs, it became an urban planning goal to counter out-migration by offering competitive ownership opportunities in the inner city. Furthermore, it was expected that the low housing prices in the historic parts of the town would enable purchasers to buy vacant buildings, and thus counteract shrinkage.

The support that was offered by the city was not a financial one. Rather, the city helped form groups of purchasers and engaged architects to give professional advice on organizational questions, project planning and costs. Together with federal subsidies, low interest rates and low sales prices, this was supposed to enable more people of various levels of income to acquire individual property in existing but vacant historical buildings. However, after a promising start, it became more and more difficult to realize this project. The main reason was a change in Germany's tax laws that led to a renewed interest from professional developers in Leipzig's historic buildings, causing an increase in sales prices and thus made the inclusion of less affluent individual purchasers more difficult. Reacting to that, and with the additional aim of siphoning demand that would otherwise be directed to the suburban fringes, the municipality expanded its Selbstnutzerprogramm towards new construction of detached and terraced houses on vacant lots. Whereas the old buildings' part of the programme came into problems, this new part enjoyed growing popularity and thus more and more individual property was realized as new construction. Altogether, as part of the Selbstnutzerprogramm, 108 apartments were thus refurbished in existing buildings between 2001 and 2006. At the same time, 237 apartments were newly constructed (Stadtforum 2006). The ratio between projects in existing but vacant buildings to those in new constructions thereby changed from 5:1 in 2001/2002 to 1:2 in 2005.

Thus, the Selbstnutzerprogramm also veered away from its original intentions, and has been the target of much critique. This critical discussion is based on three main points of contention. First, the scale of the programme was seen as marginal in the context of persistent vacancy of around 40,000 flats. Second, most projects were realized in those areas of Leipzig that have fewer problems and therefore do not need additional support. In fact, most projects are located in areas with slight gentrification tendencies where demand for owner occupation is highest. Third, the whole idea of supporting new constructions in a situation of an already existing oversupply is seen as questionable. As these new constructions are mostly carried out in a 'suburban' form, critics also argue that this policy would rather import suburbanization to the inner city than create new urban landscapes.

Conclusion

In taking stock of Leipzig's regeneration politics, the results are split. On the one hand, Leipzig's local government has indeed shown a desire for an inner city renaissance. It has engaged a vocabulary of 'cappuccino urban politics' (Peck

2005:760) and implemented instruments that aim for 'culture-led revitalization' and promote middle-class oriented 'urban idylls' (Hoskins and Tallon 2004) – a course that has become popular with municipal governments all around the world. Looking purely at the agenda for urban restructuring, Leipzig thus seems to follow the way that many other municipalities embark on worldwide. On the other hand, the reality of Leipzig's renaissance is proceeding in a rather contradictory fashion. Thousands of homes have been demolished in recent years. Furthermore, as a consequence of real estate companies' varied willingness to cooperate, the demolitions have not happened in a coordinated way but rather in an ad hoc fashion. Moreover, most intentions to embed housing market renewal into upgrading procedures have failed, or could only be realized on a minor scale. Despite intensive efforts, the result of regeneration politics can, therefore, at best be described as 'islands of renewal in a sea of decay', whose location does not even follow any particular spatial logic. At worst, it can be criticized as a devastation of compact urban form and its displacement by a random urban pattern with a reduced density.

As this outcome contradicts the original intentions, it throws an interesting light on the political economy of regeneration and shows that it is the context of particular housing markets – and not discourses and planning paradigms – that is fundamental to the success or failure of urban renaissance. If the local context is dominated by low demand, weak markets and a fragmentation of interests between local landlords, the preconditions and possibilities for regeneration change considerably. As the demolition of building structures in the short term does not create profits but costs, the interest of private capital becomes at best lukewarm.

Thus, public urban planning initiatives are very much hamstrung by the local conditions of the housing market. Under these conditions, local policies have limited options and many of those implemented prove not to be working after a short time. In places like Leipzig where neither the market can compensate for the lack of public resources, nor public resources substitute for the failure of markets, urban renaissance easily becomes an unreliable saviour and local policies are doomed to ebb between renaissance rhetoric, ad hoc action and tokenism.

9 Image politics and stagnation in the Ruhr Valley

Sebastian Müller and Constance Carr

Regional development and modernization plans have aimed to reverse economic decline, urban shrinking and social and ecological degradation in the Ruhr Valley since 1999. Internationale Bau Ausstellung (IBA), the International Building Exhibition Emscher Park, was a programme intended to rejuvenate a heavily urbanized post-Second World War industrial region in North Rhine-Westphalia, Germany – an area that had been in economic crisis since the 1970s. IBA Emscher Park was an all-encompassing ten-year urban restructuring programme that proclaimed capable of revitalizing the multitude of abandoned industrial properties, addressing the local social housing needs, and reversing environmental problems.

The socio-economic framework of the Emscher region

The location of IBA Emscher Park is part of the greater Ruhr Gebiet (Rhine-Ruhr) Area. Historically, this area was one of Germany's largest Fordist hubs of vertically organized production and manufacturing. Its recent history, however, is one typical of Western post-Fordism: deindustrialization, socio-economic decline and the arrival of flexible modes of production and services. The socio-economic framework of Emscher Park was similar to other European and American regions of primary industry such as coal mining and iron and steel production that characterized the socio-economic situation as a whole throughout the Northern Atlantic continents for several decades, and which have also been migrating to and arising in developing countries (Esser and Hirsch 1989; Lipietz 1986; Scott and Storper 1986).

The IBA Emscher Park region, however, did have its own locally specific socio-economic configurations that made its post-Fordist transformation unique among other North Atlantic experiences. First, the region was and remains primarily an urban sprawl. Today, the Emscher region has a population of roughly two million, who in large numbers commute daily to work, quality shopping or higher education facilities, which are found in the southern parts of the Ruhr Area, the Rhine belt or in the major cities of Dortmund, Bochum, Essen or Dusseldorf. Housing as well as an impressive number of motorways and power plants also dominates the landscape. Further, railway networks and waterways that were once redirected to enormous old coal and steel factories are now underused or derelict. Second, there was a sizeable decrease in the availability of employment during the 1980s and 1990s, with a shortfall particularly in the expanding fields of high technology, professional and public services and research. Unemployment rates were and remain some of the highest in Germany, with an increasing number of families depending on state transfer incomes, casual or part-time labour and/or familial support. Furthermore, a majority of the young and economically successful depart the region – at least on a daily basis – leaving behind the elderly, the socially weaker and undereducated. Third, both legal and illegal emissions from heavy industry and coal power production have led to widespread soil, water and air pollution with remarkable impacts on population and environmental health.

History of IBA Emscher Park programme and its policy model

The planning policy model of the IBA emerged out of a long tradition of Social-Democratic and Fordist restructuring politic of the Ruhr. Clearly the naissance of the IBA Emscher Park was one of the ruling political classes in North Rhine–Westphalia. More specifically, however, it was the product of a think-tank propelled by Christoph Zoepel, Minister of Urban Development, Traffic and Housing of North Rhine–Westphalia, in early 1988. The IBA policy model had four key characteristics, which will be outlined below:

1   strong public involvement;
2   strategic and incremental approach to policy-making;
3   a decentralized network-based policy model; and
4   an implementation of the programme through subsidiary discrete projects.

The think-tank had written a document, referred to as the 'IBA-Memorandum', (Ministerium für Stadtentwicklung, Wohnen und Verkehr 1988) that summarized an urban and regional analysis, and drafted guidelines for the working areas of emphasis of the IBA. The think-tank, in collaboration with the government of North Rhine–Westphalia, called upon municipal governments, industry, civic associations and the general public to propose specific projects that might be developed under the IBA programme. This was how the first unusual characteristic of IBA's planning policy model came into being, in the late 1980s. IBA demonstrated a vigorous interest in public involvement by widely publicizing its restructuring ideas for the region, and in open-handedly inviting interested parties to participate right from the beginning.

Second, IBA's motto was, 'workshop for the future of old industrial areas' (Ministerium für Stadtentwicklung, Wohnen und Verkehr 1988), which signalled a new experimental policy-making process in regional planning: a flexible approach to developing planning solutions and a learning-by-doing methodology. The Memorandum proposed programmes for Emscher Park that could be carried out by a step-by-step incremental strategy (see Ganser *et al.* 1993). This left leeway either for the later cancellation of projects that were not supported by powerful stakeholders, or for the possible modification of the agenda should support from prominent industries, municipalities, and/or instrumental media conglomerates intervene. The relative structural autonomy of the IBA from formal decision-making bodies of North Rhine–Westphalia and its municipalities also allowed for such flexibility in the agenda.

At an arm's length from the government, a small planning unit called the Emscher Park Planning Company was set up to administer the programmes on the agenda. It was a, 'wholly owned subsidiary of the Land North Rhine-Westphalia under civil law and with limited liability' (IBA 1999). It was a self-regulating body without any governmental executive power at hand, but wholly financed by the State of North Rhine–Westphalia. Furthermore, it was privileged with close relations to the ruling political party and easy access to other influential members of legislature and influential civil servants. The appointment of Karl Ganser – a leading employee from the Minister of Urban Development Traffic and Housing – as director of the IBA Emscher Park ensured this access. This independent planning unit then proceeded to set up wide-reaching networks of powerful stake-holders. Representatives from the state and municipal governments, industries, social and cultural initiatives, architectural and ecological associations, along with a healthy number of individual journalists, politicians, landscape architects, artists, designers and planners began exchanging, communicating and collaborat-ing. Thus, the third decentralizing characteristic of the IBA policy model emerged. During the ten years of the IBA's activity, approximately 400 individual projects were developed and 120 projects – spread throughout the region – were realized by contracting them between respective local authorities, the IBA and the govern-ment of North Rhine–Westphalia.

The fourth characteristic of the IBA policy model was the organization of its subsidiary projects. The IBA Emscher Park concentrated on five – and later in 1996 on six – central working areas of emphasis. The various themes were enti-tled (IBA 1999):

1   'The Green Framework: The Emscher Landscape Park';
2   'Regeneration of the Emscher River System';
3   'Working in the Park';
4   'Industrial Monuments';
5   'Housing Construction and Urban Development'; and
6   'Social Initiatives, Employment and Training'.

A few examples from each of the aforementioned six working areas are described in the following paragraphs.

The 'Emscher Landscape Park' was a patchwork of regenerated or protected green areas that were located near the Emscher River and throughout the region. The largest project was the transformation of the former steel company, Meiderich, into the 200-hectare Landscape Park Duisburg-Nor in which derelict industrial buildings and steel structures were refurbished to stand in memory of the park's industrial past, and the property on which they stood was transformed into a park. This included the regeneration and naturalization of the former industrial water systems by transforming them into ponds and canals. Other IBA greening projects included smaller bio topic installations, such as the greening of neighbourhoods, or the opening of the new Emscher Hiking and Cycling Paths.

The Regeneration of the Emscher River System programmes never did develop into the benchmark, 'symbol of ecological regeneration of an industrial region' (IBA 1999), as they were originally promoted. Some tributaries to the Emscher Canal were restored. The water quality has indeed improved, but the technical profile of the central open canal remains virtually unchanged, while the sewage is now in the process of being piped underground. There is some fragmented access to the Emscher riverbanks via the Emscher cycling path, but the central canal and sewage treatment plants that were enlarged and modernized highlight more the waste management system itself than the rehabilitation of a contaminated waterway. The river Emscher remains, just as it was in the past, invisible.

The Working in the Park projects resulted in a chain of technological centres, with the flagship project in Gelsenkirchen: the Rheinelbe Science Park. Under these initiatives, old industrial sites were converted into office space, and companies whose focus was on high-tech, media, architecture and design moved in. In addition, all of these projects had a, 'large proportion of open and green space in common, [as well as] high ecological and architectural standards' (IBA 1999). However, there are no stories of extraordinary success, nor any observations of remarkable improvement to local unemployment rates, or an economic spill-over into the region.

Of all the working themes, those projects developed under the Industrial Monuments programmes, are the most spectacular. Prominent examples include the old coal mine 'Zollverein IX' in Essen (now also designated a UNESCO World Heritage Site), and the 'Zollverein colliery' nearby. As mentioned, the Ruhr Area endured massive industrialization but very little was left of this legacy except for some bits of information filed away in archives, or remaining testimonials that tell stories of life during industrial times. The IBA projects thus aimed to serve as the region's 'monumental witness to industrial culture' (IBA 1999). Old coalmines and steel factories were sanitized and opened up as galleries, theatres, concert halls and restaurants. Today, they are hip places of (successful) industrial tourism, showcasing economic and social change and the corresponding arrival of post-industrial leisure. Yet, they remain problematic. As private investment was not particularly keen in investing into the old buildings, state authorities and municipal councils were forced to continue squabbling over shares in financial responsibility for any development and maintenance of these large sites.

The Housing Construction projects showed interesting results, such as the renewal projects Schüngelberg in Gelsenkirchen, or Teuteburgia in Herne. 'Approximately 2500 new and 3000 refurbished housing units' (IBA 1999) were funded by IBA Emscher Park. This number of units was minimal compared with the overall housing needs of the Emscher region and its two million inhabitants. Still, it must have been quite a feat to convince housing corporations to cooperate and invest in the deteriorating housing market. The large and traditional rental housing companies had already set in motion plans to change major social housing estates from a Fordist infrastructure into global commodities, and five years after the IBA's completion in 2004, Thyssen, EON and other old steel and mining companies of the region successfully sold approximately 500,000 apartments to Private Equity Capital from London – renamed German Annington. This London group, which managed the rental housing stock from a distance, all but entirely discontinued further investment towards maintenance and modernization of the units (BMVBS 2007). These homes, that were not in top physical condition to begin with, have only further deteriorated since their sale.

The working area of Social Initiatives, Employment and Training was initiated as a result of persistent pressure from civil movements, which lobbied for a response to general unemployment problems, and for support for already existing local projects, 'where energy and commitment had been initially generated by the citizens themselves' (IBA 1999). Good examples are the Riwetho housing community in Oberhausen that transformed an old company town into its own social housing association, or the 'Hands Off the City Park' in Castrop-Rauxel that took over a former public swimming pool and transformed it into a cultural and social centre. In the end, these projects managed to acquire a piece of the pie – although it arrived relatively late and small when it did.

Avant-garde position of the IBA Emscher Park?

The IBA-Emscher Park was continually applauded as a 'new innovation' (Knapp 1998:387). Danielzyk commented that IBA Emscher Park was among the 'perhaps most complex and most ambitious procedures in Western Europe' (Danielzyk 1992:101). Also positive about the IBA were Kilper and Wood, who concluded that 'the IBA [was] an innovative approach to regional development and structural policies in the Federal Republic of Germany' (Kilper and Wood 1995:217). In a similar vein, Kunzmann wrote that the IBA 'provided fresh impetus in many areas, [and] … most importantly, it created a new image in the minds of many people of a region that they had already written off' (Kunzmann 1999:79). In Kunzmann's eyes, creating this new image was the focus and perhaps the sole 'vision' of IBA:

> This vision was and continues to change the image of the region in people's minds, to create new images and to ban the old ones to the pigeonholes of stubborn prejudice. The IBA's aim was not to create jobs instantly, nor to formulate and enforce new economic and technology policies for the Ruhr area. Instead, it was to create long-term improvements in conditions, namely in sites, that are prerequisite to the area's attractiveness for investors.
>
> (Kunzmann 1999:81)

According to Kunzmann, five ideals were fulfilled:

> The first is that all projects introduced a new, hitherto unknown, architectural quality to the region … they are the flagships and signs of new times in the

Ruhr region. Secondly, the projects demonstrated that old, unused industrial buildings ... could be turned into attractive modern premises for small and medium-sized business. ... Thirdly, as a rule, the projects that were realized led to an upgrading of the location in question. The image and economic climate of the sites are very gradually beginning to improve ... because some parts of the Emscher region lack attractive sites for new economic initiatives, and not so much affordable housing for underprivileged groups of the population. ... Fourthly, many of the (business) parks are making their mark with business profiles in certain future-oriented branches; ... these parks have made statements that will not fail to show results in the medium term. ... Finally, the projects have also contributed to the fact that qualified specialists and graduates of higher educational institutions in the Ruhr area have stayed in the region.

(Kunzmann 1999:83)

Kunzmann's conclusions, however, were never grounded on research, but rather on sympathy for a neoliberal amalgam of ambitious architecture, cultural production and economic development through gentrification processes. Two quotes from the Memorandum show that the intentions of IBA, at least initially, were not as Kunzmann described: (1) 'The International Building Exhibition Emscher Park was inaugurated to push ecological, economical and social transformation of the Emscher region ... future oriented in its conceptual, practical, political, financial and organizational methods'(Ministerium für Stadtentwicklung, Wohnen und Verkehr 1988:7) and (2) 'The success [of IBA] will be tangible in a continuous improvement of living conditions, of location qualities, and of solutions for living together' (Ministerium für Stadtentwicklung, Wohnen und Verkehr 1988:10).

Nonetheless, in the second half of its existence, IBA Emscher Park actively created imagery for architects and media as a means of showcasing the success of its programmes. IBA launched the campaign, Culture of Building and Quality of Architecture in the Ruhr Area (IBA 1999), an objective that was not outlined in the original Memorandum. This developed into the production of picture books, brochures, and events, as a means to 'exhibit' the productivity of IBA Emscher Park. IBA produced and distributed hundreds of photographs of its architecture, works of art, transformed landscapes, and innovatively reused industrial structures. This collection was not modest documentary photography, as most photos displayed emotional and dramatic testimonies of the Emscher region's industrial heritage and of its transformation as a result of IBA – images which eclipsed the old, dreary realities of work and workers in heavy industry. Furthermore, these vivid images endorsed a positive sentimentality – a new 'atmosphere of objects' as Boehme (1995) put it – and constructed another post-industrial or urban reality, by merely capturing scenery for the purposes of services and leisure. The creation of these pictures opened the door for a neoliberal image campaign that advertised a new postmodern Emscher region.

In addition, the gentrification of the old inner city harbour of Duisburg, which was also carried out in this time period, was a retaking of the city for the middle- and upper-middle classes. Urban restructuring eliminated harbour and labour functions, and established high-rise office spaces – designed by the architect

Norman Forster. A new urban landscape of museums, bars and restaurants, entertainment consumption, and luxury condominiums was created. The Zeche Zollverein Schacht XII (Zollverein Pit Shaft XII) with its conversion of a former coal-mining compressor hall into a fancy restaurant, surrounded by a design museum, a design school, congress facilities and jewellery and art shops, was also a project of retaking historic sites for urban middle classes and recreating a new urban image. What characterized this second phase of IBA was the implementation of modes of 'urban regeneration' popular in the 1990s in Europe and elsewhere, described by Smith as a 'global urban strategy' (Smith 2002b:90).

Restructuring regional governance and ambivalent symbolic politics?

From its inception, IBA Emscher Park aimed to flex two aspects of regional development politics in the Ruhr Area: (1) the institutional governance methods of regional planning towards more non-governmental collaborative styles and (2) the traditional industrial top-down economic approach to spatial planning that aimed to maximize production towards more socially and ecologically integrative methods of urban restructuring. However, IBA proved to be part of a general trend in the re-scaling and reshaping of traditional territorial governance throughout Northern Europe in the 1990s, a shift which has generally been correlated with the emergence of post-Fordism in urban and regional political studies (Brenner and Theodore 2002a; Hitz *et al.* 1995).

The Emscher region itself is not a formal political unit. It is recognized neither by the state of North Rhine–Westphalia, nor by the Emscher region's constitutive municipalities. Therefore, throughout its operation, IBA was able to avert traditional administrative boundaries, while regulating an array of decentralized individual initiatives. The networking of projects allowed more informal engagement into the development process. Its flexibility legitimated bargaining, negotiation, and cooperation among diverse participating bodies (private individuals, semi-public groups and administrative boards), yet required powerful stakeholders. The outcome of IBA was a product of negotiated interests and cooperation of suitable partners who were ready to intervene at the local level of each individual project. The outcome was not a product of transparent and inclusive democratic regional decision-making systems.

The IBA policy model was pragmatic, selective and necessarily exclusive, and in effect, representation of social and ecological problems in the Emscher region ended with the IBA's Memorandum. As a result, the IBA Emscher Park overlooked and eclipsed a wide variety of regional development issues like the myriad of polluting traffic arteries and power plants, and sizeable deficiencies in public transport. Other public concerns were only marginally addressed, such as unemployment, environmental protection, poverty, integration of migrants and youth cultures of protest and difference. At best, lobby groups and activists from these fields were very selectively welcomed (at minimal levels of funding) during the second half of the programme. Yet mostly, they remained in the background, complaining of their invisibility (Initiativkreis Emscherregion 1994). IBA Emscher Park set up most of its projects through exclusive partnerships between selected architects and powerful affiliates while, if need be, bypassing the bureaucracy of elected local and regional administrations.

The partnerships governing individual projects did not alter or challenge traditional structures of power. The partnerships under the IBA policy model had significant roles as intermediaries and as facilitators. While dependent on project-based consensus, IBA produced and handled internal conflicts through innovative governance methods like professional participation, mediation or networking, but could ignore external political or ideological conflicts. To some extent, IBA Emscher Park depoliticized the discourse of urban regeneration and regional transformation. Thus, a spirit of 'muddling-through-regionalism' evolved, that was neither able nor willing to create new and effective regional forums of political debate, arrive at solutions to acute strategic contradictions, or manage efforts that sought to address the complicated social situation of a region still in decline. This was how IBA lost its roots in ecological and social movements and politics.

As a consequence, powerful industrial enterprises, city mayors or influential Social Democratic party leaders carried on with their political agenda for the region, which emphasized economic growth, raising the region's image and elevating property values. These objectives were seen by the construction of 'flagship' projects in the Emscher region, such as the private design school at Zollverein, the mega mall 'CentrO' in Oberhausen, the amusement centre of Warner Brothers film park in Bottrop, the high-tech soccer arena in Gelsenkirchen or the indoor ski hall in Bottrop (see Wirtschaftsmagazin Ruhr 2007; Initiativkreis Ruhrgebiet 2007). These were projects that restructured political and economic constellations by attracting soft skills, like tourism, culture and entertainment industries into the region (see Ministry of Economic Affairs and Energy of the State of North Rhine–Westphalia 2006). In this context, the new images of a post-Fordist better world in the Ruhr Area, as shown in IBA photographs, were indeed well received – at least, in the pockets where successful business enterprises in the fields of high-tech, media, and entertainment thrived. However, the spatial planning process placed economic profit over and above social and ecological responsibility – just as it always had.

In the end, IBA's policy model proved only a successful symbolic appeasement politic at best, and ten years after completion, most parts of the Emscher region remain in socio-economic decline. In fact, statistical indicators have shown that the Emscher region, together with the Ruhr Area, continues to differentiate for the worse from average growth rates in Germany since the mid 1970s (Boemer 1999; Bade 2006; Boeckmann 2006). A 2007 study ranking German cities along 104 indicators placed all major cities of the Emscher region at the bottom. Of 50 cities, Dortmund landed at 31, Duisburg at 32, Herne at 37 and Gelsenkirchen at 48 (WirtschaftsWoche and IW Consult 2007:30).

Conclusion

In some respects, it might be tempting to conclude that IBA Emscher Park achieved little. However, this is not our argument. Instead we want to emphasize that IBA contributed to a highly uneven geography in the Ruhr Area, where some spaces have been systematically privileged over and against others as sites of capital accumulation and well-being. How did this develop? The IBA Emscher Park Memorandum had initially adopted the popular position of critical German urban studies of the 1970s and 1980s that traditional industrial vertical development continually overlooked ecological, social and cultural problems and sought only spatial planning solutions that would improve production. IBA had set out to prove that an alternative social spatial restructuring programme was possible, and its governance and sustainable development strategy became extremely popular in the regional politics and urban studies profession. Through its discourse strategy, too, IBA developed an ideology for the region that aimed to reconstruct the increasingly fragmented agglomeration of the Emscher and the Ruhr region as a space of specific and outstanding identity.

However, through this discourse of identity and image, real and existing problems associated with the old industrial region were overshadowed, and the imagery of a postmodern leisure use of industrial space for middle and upper classes thrived. Sustainability lost out to photography, art and architecture. Ecological and social rejuvenation became subservient to economic development and growth. Restructuring politics became ever more exclusive, and planning processes were recycled and reworked into a neoliberal concept of competition – pitting neighbourhood against neighbourhood, and city against city. This concept created winners and losers. Winners were those with political and/or economic clout. Losers were those who had to consider the worsening socio-economic conditions of the Emscher and Ruhr region. There is still work to be done.

10 Gentrification and the creative class in Berlin-Kreuzberg

Ingo Bader and Martin Bialluch

Berlin-Kreuzberg has been the scene of two periods of 'urban renaissance': first, a cautious renewal policy that superseded the demolition of the old building stock for large-scale social housing projects and, second, the rediscovery of Kreuzberg SO 36, a traditionally poor district in the south eastern part of Kreuzberg, by the 'creative class' (Florida 2002) in the late 1990s as a potential new economic cluster to rebuild the Berlin economy. While urban underground cultures have played an important role in the local urban restructuring process, the new policy regime to reconstruct the Berlin-Kreuzberg area as a centre for cultural activity has overridden many of their achievements towards social mix and justice. In this chapter, we evaluate this policy shift from the locally focused, participatory process of *Behutsame Stadterneuerung* (Hämer 1990), or 'cautious urban renewal' (a redevelopment policy from the 1970s that has its roots in this district), to the economically driven and globally focused *Stadtumbau West* 'urban renewal west' policy approach used in the later project known as 'Media Spree'. In this regard, our particular focus is on the large-scale redevelopment of the banks of the river Spree and the associated neighbourhood, Kreuzberg SO 36. We argue that there are very serious problems with the new approach and we outline concerns about its impacts. However, we conclude that due to an overall bad economic performance of the city there is not much obvious gentrification so far.

Back from the periphery

The specific geography of the city of West Berlin – as an 'island' within the former German Democratic Republic surrounded by the Berlin Wall – had turned the city, and in particular the south eastern part of Kreuzberg (known as SO 36), into a peripheral zone. Kreuzberg SO 36 became a pocket extending into the East, bounded on three sides by the Berlin Wall. The river Spree had been the border between Berlin-Kreuzberg and the East German district Friedrichshain that were merged into one municipality in 2003 (see Figure 10.1).

In the post-war period, due to underinvestment and the housing stock, to some extent, still showing war damage, skilled workers started to leave the former working-class district of Kreuzberg in order to move to large modernist suburban housing estates. Starting in the 1960s, this process was increasingly accelerated

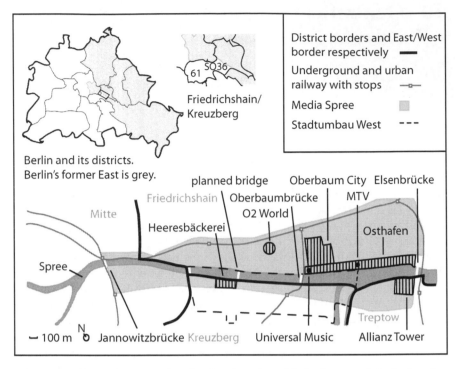

Figure 10.1 East and West Berlin, showing Kreuzberg, Media Spree, and the Stadtumbau West area.

Map by Ingo Bader.

by an extensive redevelopment plan for this area and a planned extension of the urban motorway system even though it was a dead end at this time due to the Berlin Wall. Urban planning in West Berlin was geared on a city-wide model although there were not any indications of reunification. Municipally owned and subsidized housing companies bought up the housing stock from independent landlords for demolition leading to vacancy or short-time leases. At this time, the traditional population was replaced by immigrant workers, students, radical political activists, artists, hippies and other drop-outs – the so-called Kreuzberg mix (Rada 1997:140). The district became a centre of the radical student movement that was shaking Germany in the late 1960s. The commune movement experimented with new forms of collective lifestyles, squatting and militant demonstrations appeared on the stage of political contest. The Kreuzberg mix refers not only to an ethnic and social mixture but also to a population with a partly alternative attitude and rebellious character, a strong subcultural influence and simultaneity of living, small-scale crafts and shops in the same buildings.

In the 1960s and early 1970s, the redevelopment of the area took a brutal approach involving large-scale demolition and the construction of social housing tower blocks. This was followed by the gutting of backyard buildings, which had traditionally been the locus of local economic and social activity, and, when considered necessary, the construction of new facades along the outlines of the existing block structure. In Germany, subsidized housing construction was mainly aimed at the middle-class and the rents in the new-built social housing were higher than those of the low-standard housing dating from the period of industrialization. This policy led to a housing shortage as the new construction did not keep up with the vacating of the tenants (Berger 1987).

In the late 1970s, strong neighbourhood resistance opposed the demolition policy which had been executed by a coalition of state-owned developers, local construction and real estate industries and municipal authorities. At this time Berlin-Kreuzberg was one of the centres of urban resistance and rebellious subcultures in Germany (Rada 1997). The appearance of the subculture changed. Punks replaced the hippies and the autonomous movement emerged. Neighbourhood councils – independent tenant organizations – and a strong and partly militant squatter movement developed an impressive activity in Kreuzberg. From 1980 to 1981, around 169 houses, often appointed for demolition, were squatted in West Berlin, and 80 of these were in Kreuzberg: a clear expression of these councils' political struggle (Berger 1987).

The result was the innovative Behutsame Stadterneuerung which marked a paradigmatic change in policy approach. Demolition was replaced by greater attention to social mix and preservation of Wilhelminian style housing – that important stock of housing built in the so-called *Gründerzeit*, the years of rapid industrial expansion in Germany following the foundation of the German Empire in 1871. Typical of this housing stock is a dense apartment block with side wings and rear courtyards, and sometimes several rear buildings and factories in the *hinterhöfe* (courtyards). As different classes lived in the same blocks at this time, the front buildings were often splendidly decorated with stucco. Behutsame Stadterneuerung adopted mechanisms such as rent controls (which were achieved by subsidies for the modernization of dwellings superseding a Land-guaranteed rent fixing in the old building stock) and strong tenant participation to bring about 'urban renaissance'. The main aims of this paradigmatic policy were participation, preservation of the specific character of neighbourhoods and the old building stock, neighbourhood improvement according to the needs of the inhabitants, gradual renewal of buildings, and solid financial support for this policy (Hämer 2007).

In 1989, with the fall of the Berlin Wall, everything changed overnight. Kreuzberg suddenly moved from the fringe to the centre of the city, and the dead-end streets at the border became important connection roads once more. Although the gates were opened to a typical mode of capitalist urban development, the gentrification of Kreuzberg (and possible loss of its distinct and unique features) was delayed for several years. The city leaders' visions of a global city did not create a booming real estate market, but instead led Berlin into bankruptcy (Scharenberg 2000). In addition, the city's focus on urban renewal in East Berlin postponed changes in the western part of the city (Bernt 2003).

The industrial wastelands on both sides of the river Spree came to be used by subcultural actors. After the grim and disaffected Cold War in 1980s, parts of West Berlin's subculture shifted to a more hedonistic lifestyle in the years after reunification. Dancing replaced demonstrations, and extensive vacancy and the absence of governmental control during the transition period formed a unique, innovative and, in the beginning, illegal club culture. Some of the places that branded Berlin the 'tekkno city' (Scharenberg and Bader 2005:7) and eventually laid the foundation for the city to become a central node of global cultural economy grew in this area (Bader 2004). A number of small-scale, locally grown enterprises of the cultural industry, as well as Berlin's hippest club, bar and live music scene, have been flourishing in SO 36 and Friedrichshain near the river.

A new vision for the area adopting the creative city concept (Florida 2002), aimed at promoting cultural industries, was catalysed by the relocation of Universal Music Group Germany in 2002 and MTV Central Europe in 2004 into old warehouses of the Osthafen (East Harbour). Both organizations claimed that Berlin's alternative culture was an important factor in their relocation (Bader 2004). In fact, the small-scale music economy cluster around Universal is, in contrast to the national branches of the major companies of the global musical industry, genuinely global in respect to markets and perception. The major branches sell what the predominantly Anglo-American market produces; national acts rarely succeed at a global level. The city's electronic music scene, the independent labels like Kitty Yo, Tresor Records, Kanzleramt and the city's club culture are accountable for Berlin's reputation as a global music city. This creative environment is used as a brand in the Media Spree development and is seen by the government as an important economic activity. Furthermore, the independent labels that were originally founded as counter-models to the global music industry in the 1970s and 1980s have revitalized their former rivals. The flexible integration of independent labels into the major music companies, whether directly as sub-labels, by using their distribution channels, or as formally still independent, as 'creative laboratories', has promoted a reorganization of the music economy that fostered Berlin's rise to global music city (Scharenberg and Bader 2005).

This particular area on the river banks, now dominated by international real estate interests and large-scale landowners, has been branded 'Media Spree' by the responsible redevelopment agency, which also operates under this name. The redevelopment plan, entitled 'Stadtumbau West-Spreeufer Kreuzberg' (Urban Renewal West-Spree Banks Kreuzberg) intends to create an urban environment for a creative cluster by regenerating the Wilhelminian-style buildings and reinvigorating the cultural economy. In the next section, we describe the Media Spree/Stadtumbau West approach to urban renewal.

Media Spree – Berlin Harbour City

Media Spree started in 2002 as a private sector marketing company and now represents 19 large real estate companies and property owners, including local and federal government-owned companies. In 2005, Media Spree was transformed into the non-profit company Media Spree Regionalmanagement e.V. (e.V.: *eingetragener Verein* refers to the status by law, roughly incorporated society). By law, the company has to execute public interests, and therefore does not have to pay taxes. The advisory body of Media Spree consists of members of the district, the Senate of the Land (State) Berlin, the local job centre and the chamber of commerce. Its jurisdiction is a 3.7 kilometres long and 180 hectares large area (eight times larger than Potsdamer Platz, the largest redevelopment project since reunification) on both sides of the river Spree, between the bridges Jannowitzbrücke and Elsenbrücke. This new quarter, Berlin's largest waterfront development project (Meyer 2006), draws the city centre eastward.

The anchor projects are the O2-World entertainment hall for 17,000 spectators (built by the Anschutz Entertainment Group) and the Osthafen with MTV Middle Europe and Universal Music Germany. The Senatsverwaltung für Stadtentwicklung (Senate Department for Urban Development) assesses the potential of this area at 2.5 million square metres gross floor area of office space housing 40,000 employees, but Media Spree's reports on recently established companies refer mainly to businesses in the lower service sector (Media Spree 2007). Beyond these few key projects, there is a large gap between the extent of the intended projects presented in Media Spree's glossy marketing magazines, and projects which actually have a scheduled construction start date.

Recently, a feasibility study for Stadtumbau West–Spreeufer Kreuzberg (Büro Herwarth and Holz 2006) was introduced by the Senate Department of Urban Development to address the development of infrastructure and public space in the Media Spree area and the adjacent parts of Kreuzberg. This legal planning document, financed by the federal programme but applied by the Land Berlin provides a framework for the redevelopment process (in Germany urban planning is the task of the municipality, but as Berlin is a Land (State) and a city simultaneously, the Senate has almost all the administrational powers of a municipality and a Land). The main objectives are translated as follows:

- 'to grasp the opportunity to redevelop part of the city with a new economy that will benefit the wider metropolitan area;
- to integrate the Spree area with the wider urban structure;
- to connect the district to neighbouring disadvantaged areas to provide the opportunity for a sustainable redevelopment;
- avoid uncoordinated developments between owners, investors and users' (Büro Herwarth and Holz 2006:9).

In particular, creative industries are seen as the mechanism to regenerate what is perceived as a derelict urban area, as shown in this quote:

Looking at the implementation strategy, the club scene and new established firms of the music economy demonstrated in an exemplary manner how a vital urban culture can establish itself on urban wasteland.

(Senatsverwaltung für Stadtentwicklung 2007)

The new planning act adopted to deliver Stadtumbau West policy as well as the interests of the Media Spree's developers, however, required the removal of

Figure 10.2 Berlin creative spaces. Empty site at Schlesische Straße/Cuvrystraße with street art at the fire wall of a typical backyard factory now housing creative industries. A plan for the hotel and office complex Neue Spreespeicher (New Spree-Warehouse) has been scheduled for ten years.

Photograph by Ingo Bader.

many of the crucial creative uses in this area to make way for new large-scale development. Some of the clubs have been forced to leave, and the beach bars' leasing contracts are vacant or were annulled already. Thus, these original creative communities have been used as an interim tool for upgrading (Lange 2007). A particularly clear demonstration of this is the Heeresbäckerei, a listed industrial monument (it was a bakery of the Prussian army), where the developer cooperated with a temporary and now closed art project to brand the project as a hip loft location (Springer 2006). In the centre of Berlin, 'cultural manufacturers are now tolerated only as interim users' (Krätke and Borst 2000:154) and the orientation of renewal policy towards the interests of the real estate industry undermines the formation of efficient clusters of the cultural economy (Bader 2004; Krätke 2004). The short-term use of small locally grown cultural entrepreneurs for cultural branding, and then subsequent marginalization, is detrimental to the cultural life of the city. They also need long-term prospects (Shaw 2005b).

Due to the reduced demand for industrial space in Berlin, however, and the consequent delay of many real estate-driven development projects in this area, the pressure on small-scale cultural users is currently not very high (see Figure 10.2). If they are forced to abandon their current locations, there is still more vacant space, though this is more and more on the outskirts. Reducing interim and small-scale users to being solely a marketing tool for real estate in the city, combined with the lack of any strategy for their support, undermines the development of a proper long-term creative city policy – a failure with dreary prospects when large-scale development projects regain strength again.

Shifts in the aims of revitalization

The aims of the different revitalization policies in Kreuzberg can be distinguished according to scale, scope and the role for public participation. Here, we set out firstly the original Behutsame Stadterneuerung approach, and then contrast this with the new Media Spree/Stadtumbau West approach.

Behutsame Stadterneuerung was aimed at tenants and improving their living conditions and the preservation of neighbourhoods was an essential component in achieving this objective. The spatial scope of the strategy was the block and neighbourhood level. There was considerable attention paid to the urban structure and morphology of the existing buildings and established patterns, in particular the Wilhelminian-style block structure. There was also much attention given to the social structures of the local people, their concrete interests, and their preferred way to occupy space. In Berlin-Kreuzberg during the 1970s and 1980s, many community-based and self-organized infrastructures were developed by the inhabitants: neighbourhood gardens in the backyards, community centres, alternative bookshops and kinderläden (anti-authoritarian kindergartens). Thus, the process relied heavily on the participation of residents. The mix of users of different social groups, ethnicities, minorities, classes and local business (including craftsmen and old-established producing industries) in one area and building complex was adopted in the city's planning concepts as the Kreuzberg mix. The integration of the demands of the residents as well as the established local economy was secured by the strong position of neighbourhood organizations in the renewal decision-making process. Organizations were often deeply embedded in the squatter movement of this time. The renewal process was controlled by the tenants' self-organization, and the new public–private actors. The established social structure was preserved by a de facto rent control, public funds, the limitation of housing upgrades, and the declaration of *Sanierungsgebieten* (preservation areas) (Hämer 1984).

In the 1980s, the local state made a shift away from the demolitions of the 1960s and 1970s towards an urban renewal policy that ensured the implementation of Behutsame Stadterneuerung. This shift brought a new approach to decision-making processes. Government actors were weakened in the new urban regime and new actors, including the Internationale Bauausstellung GmbH (IBA – International Building Exhibition company, with limited liability) and STERN Gesellschaft für Behutsame Stadterneuerung (Company for Cautious Urban Renewal) were established in their place. These companies effectively institutionalized the civic actors who supported the protests against the earlier demolition policy. The IBA GmbH was founded in 1978 to prepare the International Building Exhibition in Berlin from 1979 to 1987, which achieved acceptance of Behutsame Stadterneuerung within the German planning community. The IBA is the leading German architecture and urban planning exhibition initiated by the

Lands to introduce social reform models to contemporary urban design. In 1985, IBA GmbH was followed by STERN, which remains a key actor in Berlin's redevelopment policy, responsible for implementation of urban redevelopment projects and building renovation in Berlin, with a fiduciary duty to the municipality (Hoffmann-Axthelm 1987).

An important rudiment of this paradigmatic change in redevelopment policy was the institutionalization of residents' participation, along with the shift from demolition to preservation. Unlike IBA GmbH and STERN, the tenant and neighbourhood organizations like SO 36 e.V. represented exclusively the political and individual interests of the population of SO 36. They gave legal advice and organized grass-root protests (including squatting) as well as the direct participation of the tenants in nearby social housing blocks. They also established a link between those local people and the fiduciary actors of the Land Berlin. The new form of organization broke down the strong traditional alliance between local state and industry, and strengthened civic actors and grass-roots movements in the regeneration processes. They represented not only a new institutional form, but integrated different forms of struggle in one regime.

Since the late 1980s, however, the financial support for these actors has been severely reduced. As a result, citizens' action groups were driven out, and only professional companies and other large organizations are now able to participate. This development was predicted by a few admonishers. Their criticism of formalization and institutionalization of the participation process and the introduction of private actors to the urban renewal policy turned out to be far sighted in light of the contemporary neoliberal domination (Homuth 1984). Even though the regime of Behutsame Stadterneuerung certainly served poor people's interests to a great extent when it was introduced, it could be interpreted as a precursor of a neoliberal urban policy because it established private actors in the urban renewal process (Bernt 2003).

Formally, Media Spree Regionalmanagement e.V. does not differ from the actors of the Behutsame Stadterneuerung as it is a non-profit organization. Media Spree's main function is the marketing of the real estate and the support of a 'creative city' planning process. However, participation involves mainly the large economic actors and investors (not producers). Grass-roots movements only have limited influence due to some support from the left-wing local district government, which is the forum through which civic actors can now be involved.

Yet despite their marginalization, some grass-roots movements continue to influence the planning process. In September 2007, for example, a *bürgerbegehren* (public petition on a formal governmental level) was started by a grass-roots organization called Media Spree Versenken (Sink Media Spree) to alter the development scheme. Due to its success, strong pressure on the local government has already been generated. Media Spree Versenken originates from the tradition of the social struggle in Kreuzberg.

Media Spree and Stadtumbau West have as their wider targets the performance of the metropolitan economy. The focus is on how this particular locality can form a major economic cluster to contribute to that performance. The model for urban development shall 'convey dynamism, the end of stagnation, the inauguration of change' (Büro Herwarth and Holz 2006:10) with the objective to 'establish a new creative mix in Kreuzberg' (ibid.:16). This appears to maintain a commitment to the concept of 'mix' as it existed in the cautious urban renewal approach. However, the value of the new 'mix' is defined by economic potency and real estate value, rather than the contribution of established residents of the area. The new policy's goal is not to improve the living conditions and economy of existing residents, but to attract the middle-class. Replacement is not explicitly mentioned, but the conceptual upgrading of the neighbourhood and its social structure is obvious. The Anschutz Hall, the largest entertainment hall in Berlin, by way of example, is a project with a metropolitan-wide scope. The main projects of Media Spree are of a magnitude that can only be exploited by major companies, not by the small but successful firms that are established in the courtyards of Kreuzberg.

At first glance, the commitment of the Behutsame Stadterneuerung approach to the existing physical urban structure and industrial heritage appears to persist in the new policy regime as the preservation of the old industrial sites on the river banks (often listed monuments of industrial history) represents an important part of Media Spree's branding strategy. These buildings are interesting objects for loft developments, though the architectural layout is, from the viewpoint of preservation law, sometimes questionable. But the new approach reduces the intent of preservation to a mere protection of the old physical stock, and removes the essential component of social participation and attention to local people and their activities. For example, courtyard factories are the preferred locations of the entrepreneurs of the cultural economy, but this is not recognized in the new regime, under which courtyard buildings are again being cleared. Participation in the new regime has degenerated to mere monitoring of the decision process (Hoffmann-Axthelm 1987; Hämer 1990), and civic actors find it difficult to be involved:

> The poor will always be the victims of good intentions. Why don't we listen to them to find out what they want? What matters is to activate the social potential in an area and to help the residents to find solutions, not to solve the problems for them.
>
> (Hämer 2007)

Whilst the feasibility study of Stadtumbau West (Büro Herwarth and Holz 2006) points out the 'high proportion of socially disadvantaged sections of the population', no measures have been established to ensure that such social disadvantage is justly addressed. Rent controls stipulated in the 1980s are now expiring, and the rents are quickly realigning to market rates due to the present system of *ortsübliche Vergleichsmieten* (comparative rent) to regulate the housing market.

Conclusion: what are the impacts on gentrification?

Compared with other capitals and world cities, there is relatively little gentrification in Kreuzberg. A significant change in the neighbourhood is visible in relation to the commercial uses, but no displacement is yet discernible. The slow and so far incomplete gentrification process can mainly be attributed to the overall weak economic performance of Berlin. The most recent renewal policy (Media Spree/Stadtumbau West) seeks to enhance that economic performance to bring about the changes that we would classify as gentrification, through minimizing regulation of the real estate market on a local level and shifting from local participation to real estate-driven urban renewal. The shifts can be summarized as follows:

1   Civic and neighbourhood participation is superseded by public–private partnership;
2   The new target group is the middle-class, the new 'creative mix', not the poor and long-established neighbourhood residents;
3   A strong participatory institutional planning framework has been undermined and replaced with the mere monitoring of developer's activities;
4   The policy focus has changed from the existing neighbourhood to economic cluster building; and
5   The subculture is in transition from a social movement to a brand.

The severe economic difficulties of the city find their expression in this shift in policy. Budget restrictions due to the city's bankruptcy let the government withdraw from heavily subsidized policies. A high unemployment rate and low economic activity force Berlin into an intensive competition policy to attract future growth-oriented developments. Using the local (sub)culture as a pivotal marketing instrument, a policy of upgrading and strong support for large-scale projects of the private sector can be seen as an urban renaissance policy aiming to strengthen an important economic cluster. Yet the local state's orientation towards large-scale projects, and towards land utilization determined by attracting those who will pay the highest rents rather than by local political decisions, undermines the very conditions important for the 'culturepreneurs' (Davies and Ford 1999:9–11) who provide the foundation of the local cultural economy cluster, and represent one of the few urban industries of global importance in the city's economy. Even if we accept that a policy of job creation through economic growth is an indispensable component of a socially just urban regeneration policy, the new policy framework fundamentally fails in this regard as it does not meet the demands of the cultural economy – the original impetus for Berlin to become a global media city. The Land and district is undermining, in Berlin-Kreuzberg/Friedrichshain, what it seeks to create: jobs, wealth and value in a previously marginal space.

Acknowledgement: Thanks to Paul Legato for his support.

Part III
On grass-roots struggles

Gathering memories at the battlefront: www.oldbeijing.org

Yi Jing, Giovanni Allegretti and James McKay

A story of resident and activist struggles against the demolition and gentrification of Beijing's historic centre.

Narrow alleyways, *hútòng*s, running between courtyard homes, *sìhéyuàn*s, have formed the heart of the city of Beijing since Mongol times, housing generations of merchants, revolutionaries, poets and others.

During the past century's Great Leap Forward and Cultural Revolution, many of these 'symbols of feudal decadence' were claimed by the state and allocated for work unit accommodation.

This century, the few *hútòng*s that remain are being destroyed to make way for generic westernized high-end retail/office/leisure/residential sprawl.

Evictions and demolitions are quick events. There is little or no time to discuss compensation, or even where to go.

Developers maintain robust working relationships with government officials, and the legal system is too weak to protect the interests of local residents.

Official protection is granted to temples, graveyards and imperial palaces, but the traditional urban fabric – poor, cranky and unkempt – is 'sub-habitat'.

Demolitions are quick events. Economic miracles do not wait.

The records show more than 3,600 *hútòng*s in the 1940s. There were only 1,200 in the year 2000, 700 in 2005, and 500 in 2006.

According to the Ministry of Domestic Affairs in 2005, there were 87,000 'public riots', typically directed against episodes of corruption, illegal expropriation and violent evictions.

Numbers are never straightforward. Official statistics say 6,700 households have been evicted from old Beijing. International observers say 1.5 million households.

This is a country where dissent is invisible in the media, and organized demonstrations are illegal. There are attempts at resistance, but they are shy and fragmented.

One activist evicted from the quadrangles of central Beijing, Zhang Wei, carried with him to his new home a 100-year-old wooden window frame. His urban memory Web site www.oldbeijing.org became the focus for a group that collects and publishes less concrete reminders of the past: images, maps and stories of the old *hútòng*s.

There are other agencies that educate residents about their legal rights and the latest developments, like the newly passed property law. Volunteers bicycle around the remaining *hútòng*s, monitoring the implementation of the city's conservation laws, collecting reliable information on the ground.

Small victories open windows of hope. There is increasing sensitivity to the fate of the *hútòng*s in the corridors of power. Some areas have been saved from demolition.

In the *hútòng*s nearest to the city's major attractions, small tourist businesses are thriving, and there are restaurants and bookshops.

The physical fabric of these neighbourhoods is safe, but even when the *hútòng*s themselves are preserved, the inhabitants remain at risk. Now foreigners and other incomers are buying up the old quadrangles, and land values are beginning to rise beyond the reach of established communities.

Shy, fragmented attempts at struggle are entering a phase of networking and interconnection.

Is it possible to sustain and grow historic urban communities in the face of a double threat: developers and their demolition crews on the one hand, displacement from gentrification on the other?

Small victories open windows of hope.

Data in this poem were sourced from *China Daily* (16 May 2007: 10), Rampini (2007) and Amnesty International (2007).

Hútòngs in Beijing, by Liu Yangfei. *Bicycles pass a shop in central Beijing*, by Jiang Feng. *Birdcage, Beijing*, by Yi Jing. *Residents of the hútòngs, Beijing*, by Wang Long. *Single flower, Beijing*, by Yi Jing. *Residential life in the hútòngs 1, Beijing*, by Yi Jing. *Traditional and modern 1, Beijing*, by Zhang Wei. *Mega-development, Beijing*, by Zhang Wei. *Traditional and modern 2, Beijing*, by Jiang Feng. *Woman and graffiti, Beijing*, by Yi Jing. *Music in the demolition, Beijing*, by Zhang Wei. *Recording demolitions in Beijing*, by Wang Long. *Residential life in the hútòngs 2, Beijing*, by Yi Jing. *Small business folk in the hútòngs, Beijing*, Bai Hao. *Residential life in the hútòngs 3, Beijing*, by Zhang Wei. *Old man, Beijing*, by Zhang Wei. *Children in hútòngs, Beijing*, by Wang Long.

11 The contested reinvention of inner city Green Bay, Wisconsin

Marcelo Cruz

Broadway Street is the main commercial street connecting the inner city neighbourhoods of Fort Howard and Seymour in the city centre of Green Bay. In the past ten years, the district around Broadway Street has undergone a significant transformation. The local community in 1994 had among the lowest household incomes in Green Bay. The average household income for the whole city was US$ 36,000, compared with $16,000 for the district. Approximately 30 per cent of the population were children 18 years of age and younger, and a significant proportion (approximately 18 per cent) were elderly on fixed incomes. Thus, almost half the population, 48 per cent, did not produce incomes either because they were too young to work or were retired. Many of the households were female headed and participating in the lower end of the labour market. These demographics help to explain, in part, the poverty in the Broadway district.

Structural changes in local urban politics and economics contributed further to this relative poverty. For the past 30 years, the middle classes have been leaving the inner city for the new suburbs of Green Bay, lowering the tax base of the city centre. Between 1970 and 1990, 83 per cent of the industries closed down and moved out of the downtown, setting off a rapid decline in good-paying union jobs. The loss of salaried jobs and the shrinking tax base as middle-income groups continued to move out to the suburbs added to the economic stress of the neighbourhoods of the Broadway district.

Broadway Street was also badly affected. Many small businesses closed and the strip was characterized by empty storefronts and low-end bars which became problematic for law enforcement. The declining appearance of the district added to the perception of the district as crime infested, full of alcoholics, 'drunken Indians', 'white trash' and unwanted minorities. The condition of local housing became an issue as well, with many absentee landlords neglecting their properties or dividing single-family homes into duplexes. The decline of investment from public as well as private sources led to the district being considered 'blighted'.

Figure 11.1 locates the district and its relation to the rest of the city centre. In recent years, the City of Green Bay has attempted to revitalize and rejuvenate the central core, identifying it in its Green Bay Smart Growth Comprehensive Plan (City of Green Bay 2003). The Fox River divides Green Bay into east and west sides. The river was essential in the development of the industrial city, which

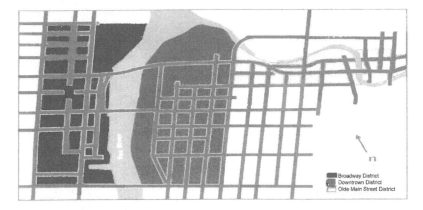

Figure 11.1 Map of Broadway and downtown districts, Green Bay, Wisconsin.
Map by On Broadway Inc.

revolved around the paper and food processing. It was a working river and became one of the most polluted rivers in the country. The exodus of industry has prompted a major environmental clean up of the river, and the Comprehensive Plan proposes using the river as an element to unite east and west. The focus on the river brings the Broadway district into the centre of the central city plan. The following is the story of how this came to be.

The beginnings 1995–1998

In October 1995, a coalition of residents and local merchants began to organize and seek alternatives to the unresponsive local government that had so let their neighbourhood decay and decline in both physical and social terms. They looked to outside sources and found a state-wide programme, the Wisconsin Main Street Program, that was set up to renovate small rural centres in Wisconsin. The programme was not designed for large urban settings like Green Bay. Nevertheless, there was much enthusiasm from the citizen group. A letter-writing campaign was organized to mobilize the residents and local merchants of the district and to attract the attention of a new generation of political leaders in City Hall.

It is important to identify the various groups and interests in this initial grass-roots coalition. One group was made up of citizens concerned about the degradation and negative perception of their neighbourhoods. The most vocal and prominent of these was based in the Fort Howard Neighborhood Resource Center. The Center is located in the Fort Howard Elementary School just west of Broadway Street. It was the mothers who volunteered at the centre who mobilized the neighbourhood through the letter-writing campaign. Their main concern was for the safety and health of the children that had to traverse the district riddled with bars and 'unsavoury types' when going to and from school everyday. They complained that their grade-school age children would emulate what they saw in adult behaviour in the neighbourhood by playing drunk. They were also tired of being perceived as 'welfare moms' and burdens on society. Getting involved in this project brought self-esteem to the mothers and a demand for respect from the rest of the Green Bay community.

The second group was the Broadway Street merchants. This group was instrumental in ensuring that the local city officials were aware of the needs of the commercial district. The merchants saw regeneration as an opportunity to enhance their businesses and attract new customers from the wider Green Bay region. Many blamed not only the physical deterioration of the district, but the social make-up of the neighbourhood, for intimidating potential middle-class customers from the district and more importantly their stores and shops.

The two main groups in this coalition that became On Broadway Inc. had inherent tensions and contradictions right from the start. On the one hand, the small but vocal residents group, tightly organized through the Fort Howard Neighborhood Resource Center, wanted to improve the quality of life for themselves and their neighbourhood. This included issues of health and safety and lowering the crime rate, which was perceived as being rooted in the many taverns on Broadway. The merchants, on the other hand, saw regeneration as an opportunity to improve their businesses and to expand their market beyond the neighbourhood. The tensions derived from different geographies: the residents were preoccupied with life spaces and improving the quality of life of the neighbourhood; the merchants defined their geography beyond the life spaces of the neighbourhood to the broader markets of Green Bay. This would bring more traffic to the neighbourhood and, from the residents' perspective, increase congestion and pollution.

A petition for the Green Bay city centre to be considered part of the state's Main Street Program was presented to the local politicians at City Hall. This was done to notify the mayor's office of the intention of this incipient grass-root organization to appeal to the state government for aid in revitalizing the commercial street of the neighbourhoods in the Near West side.

The new mayor in City Hall had witnessed the decline of the city centre and wished to reverse the trend. The last comprehensive plan for the city had been prepared in the 1970s and it was in need of updating. The coalition of residents and merchants was joined by City Hall in the push for Main Street Designation by the State of Wisconsin. With city support, members of the neighbourhood group, local merchants and city officials travelled to Madison to request this inner city neighbourhood to be a part of the Wisconsin Main Street Program.

The state was reluctant to designate the Broadway district a Main Street Program because the programme was not designed for large urban areas. But history was made when the delegation convinced state officials that the district would be an innovative use of the programme. The solidarity shown among the residents, merchants and politicians was key in their success. The coherent and positive nature of the group was clear: the neighbourhood associations were able to improve the quality of life for their families and restore the integrity of their neighbourhood, and their case now received attention and respect. The merchants saw that by improving the image and reversing the physical deterioration of the commercial street, their objective of tapping into broader local and regional markets could be achieved. The administration in City Hall saw the opportunity to increase the city's tax base by attracting new businesses and retaining older ones. Local officials recognized the importance of Broadway to the plan to regenerate the entire city core. This show of civic mobilization, from a part of the city that had been written off by local officials and in popular perception, was considered extraordinary indeed.

There was great enthusiasm and hope among the residents. Three town meetings were held during 1995 and 1996 to create a mission statement that had the goal of regenerating the commercial strip while maintaining the physical and social 'integrity' of the Broadway neighbourhoods. These town meetings drew large crowds and produced a statement of a vision for the district. The statement was sensitive to the need to change the image of the district without producing the negative effects of gentrification, which was a main concern of the local residents.

The first director of On Broadway Inc., with technical advice from the state, set up local committees on organization, promotion, design and economic restructuring, all of which were overseen by a board of directors. The board was also drawn from the three factions of residents, merchants and public officials. The board and committees were crucial in directing the policies and thinking over the next three years, from 1995 to 1998. The physical infrastructure improvements were planned and implemented during this period. A new streetscape design was initiated for Broadway Street, with further funds from the city. The design provided for wider streets and a pedestrian-friendly environment. Older structures would be salvaged and renovated through a historic preservation programme. Street lighting would illuminate the buildings from the sidewalks and trees. New infill buildings were also planned and designed so that they would not break the urban fabric of the older commercial district. The design of the district's revitalization kept true to the mission of maintaining the neighbourhood's physical integrity (see Figure 11.2).

The social element was more contentious. One of the first issues to reveal the underlying tensions was the homeless shelter located on the northern side of Broadway Street. Certain members of the committees and board, particularly the merchants, wanted to close and move the shelter, preferably out of the district altogether. Other members insisted that the homeless shelter was a part of the neighbourhood and that the residents were their neighbours as well. The debate also split the residents. The shelter was allowed to stay on the proviso that it would eventually find a new location. The debate exposed a growing distinction between two visions for the district: one, led by important local merchants and some public officials, to 'clean up' Broadway, with an emphasis on up-scaling businesses to attract higher income groups to the neighbourhood, and one that viewed with alarm this trend that would impact negatively on its social integrity. The local residents were vocal in their argument that the vision statement should address the importance of maintaining the working-class nature and historic links to the industries of the Near West side.

Figure 11.2 Revitalization of Broadway Street, Green Bay, Wisconsin.
Photograph by David Guba.

Transition phase 1999–2002

The period between 1999 and 2002 was one of transition for On Broadway Inc.
A new director came in with the intent of making the grass-roots organization
'self-sustaining'. Funding of the organization had become an issue. Wisconsin's
Main Street Program provided technical and funding support for the first three
years and then it was up to the organization to remain viable and sustainable.
Further funding would have to come from the city. Maintaining good relations
with the city was therefore seen as vital for On Broadway Inc. to survive.
Promotion and fund raising dominated the period.

A development that went unnoticed was a subtle tip in the balance in the board
of directors in favour of the merchants who advocated up-scale food establishments and boutiques. Meetings had always been held during the day or early
evening, making it difficult for working-class families to participate. This group
of merchants began to court and find allies in the local politicians. This strengthening alliance was crucial to the second and third phases of the streetscape
improvement, which was by then entirely dependent on city funds and resources.
The funds came directly through the designation of the district as a TIF (Tax
Incremental Financing) district, which allowed tax revenue generated from the
improvements to be reinvested in the district.

New developers, attracted to the district by this funding policy, did construct
new buildings along the commercial strip, and began to renovate older buildings,
mostly factories and warehouses. An old cheese factory was converted into a

restaurant, and was recognized by the Main Street Program for best reuse of an existing building in 1999. These improvements were recognized as tangible evidence of 'progress', and the district was indeed being 'cleaned up'. The new businesses were definitely up-scale and some of the residents became quite concerned. The improvements became alienating to those lower-income residents who missed the 50 cent tapper of beer at the local pub, and could not afford five dollars for a pint of imported beer. The newer eating establishments were clearly meant for an upwardly mobile clientele. A growing number of residents began to resent the institution that had initially given them so much hope. There was a sense of a loss of ownership of the street, of the organization, and of the process.

The board and committees became dominated by merchants and local politicians wanting to attract the middle- and upper-middle classes to the city centre. As the Main Street Program decreed that board membership was by invitation, new appointments consolidated the majority and were made on the basis of access to City Hall. The residents recognized that money needed to be brought into the district, but most of the new projects and businesses were beyond their economic reach. At the same time, they lost their grocery store and many other establishments that had catered to the lower-income residents. The dynamic of the organization during this period became one of conflict.

Maturing phase 2003–2007

From 2003 to 2007, a third director oversaw the maturing organization. This period saw its institutionalization. Much of the initial enthusiasm from the residents had transformed to cautious optimism and finally to cynical resignation. The board was solidly controlled by the local merchants. City Hall had a new mayor, and although he expressed a desire to continue the work of his predecessor in revitalizing the city centre, his administration was more pro-business than the previous administration. The city invested in major infrastructure improvements as Business Improvement District funds became available. Broadway Street was designated an entertainment district, without the question being asked to whom it would cater. Liecht Memorial Park, a green space on the waterfront, officially opened in 2006 as not only green space for the local residents but to host major events such as Bayfest, Tall Ships, Mexico Independence Day festivities and Green Bay's fourth of July celebrations. These events drew people from the larger Green Bay metropolitan area to the district: the regeneration was a success.

In 2005, On Broadway Inc. celebrated ten years as a non-profit grass-roots organization. There was a feeling among both residents and merchants that it was time to revisit the mission statement. The third director concurred and a series of town meetings were held. A new vision statement was drawn up with the express objective of creating a live/work environment for the district. This was interpreted by the local residents to mean the creation of an urban environment that provided the possibility of walking to work and to the shops. For the local merchants and local politicians, it meant creating a middle-class clientele and a neighbourhood that would attract young professionals and middle-class families to live in the city centre. The new vision statement, albeit vague, led to the development of a new Comprehensive Plan specifically for the district. This was another first for a Wisconsin Main Street Program.

The Plan addressed the physical and social dimensions of the district's regeneration. It proposed mixed-use development that would bring the middle classes to the inner city, and ensure affordable housing for existing local residents. It also proposed that new businesses in the district should cater not just for upwardly mobile young professionals but also working-class families residing in the district. The difficulty was how to create incentives for businesses to cater to low-income working-class residents. This was not fully addressed in the Plan. To be fair, however, the Plan itself is a strategic document which would be developed for implementation. It is important that the document began to address issues in the district that had not even been articulated in previous years.

The new Comprehensive Plan is an attempt to counter the prevailing gentrification of the district. It set up design guidelines that, although not binding for developers, made clear the role of On Broadway Inc. as an advocate for live/work environments that maintain the physical and social integrity of the existing neighbourhood. Of course, the vagueness of the mission statement leads to different and often contradictory interpretations of what it actually means. This will likely exacerbate tensions between merchants, politicians and community organizations.

The Plan promotes mixed-use, high-density development with links to the neighbourhoods of the Near West side. It identifies new development integrated with existing activities, and facilitates the links between civic spaces (public spaces), residential neighbourhoods and commercial areas. It goes a step further than the previous plan in addressing the needs of the existing residents.

On Broadway Inc. has matured in its 12 years of existence. Its continuing redefinition has been borne out of contest between the different groups that made up its board and committees. The organization has had three directors and an interim director. The role of the directorship was critical in trying to negotiate the contrasting visions and often conflicting views of what the organization should be doing for the district.

Since 1995, 54 new businesses have opened in the Broadway District, creating a net of 664 jobs. Most of these are service sector jobs in retailing, food and specialized services. Over US$ 38 million has been privately invested in the district. In the same period, the district has lost its grocery store, a pharmacy, the neighbourhood theatre house and local eateries.

In 2007, On Broadway Inc. purchased a large site in northern Broadway Street. This is another first for a Main Street Program, in that a non-profit organization will oversee a major redevelopment project on land it owns. The renewed challenge is to implement the Comprehensive Plan in a way that allows the neighbourhoods to have a voice. The district has changed its perception of being a 'slum' and 'blighted' area; however, the median income of the neighbourhoods that make up the district are still below the city's, and the gap between the city's affluent and the poor seems to be widening. The elderly population is being replaced with more upwardly mobile professionals, but there are still a significant proportion of youths that have no space designed for them and few job opportunities. More adults are working in the district but the wages are not increasing, adding to the number of working poor in the district.

These are the challenges to On Broadway Inc. as it enters its 13th year. There is still promise and excitement in the organization, but can it maintain its autonomy from City Hall and meet the needs of both the upwardly mobile professional and the working poor? Can it create a balanced policy that addresses the needs and aspirations of residents and merchants? Can it maintain the physical and social integrity of the community and still create a built environment that provides work and live spaces for *all of its residents?* This story can only be understood as a process of democratic participation that is riddled with tensions, but from that, creativity emerges. Only mediocrity emerges out of forced consensus and the absence of dialogue. Community should not be seen as a given but rather as a process that is in flux and dynamic and the community is continuously being contested and restructured in space. The answers to these questions are blowing in the winds of uncertainty and fluidity that is the urban condition in today's contemporary city.

12 Planning from below in Barcelona

Marc Martí-Costa and Jordi Bonet-Martí

This chapter looks at the potential and limits of social mobilization to defend an industrial site in the neighbourhood of Poblenou – a central part of Barcelona's urban plan. The case of Can Ricart provides valuable insight into the ways various, seemingly distinct, actors can work together to redefine and participate more effectively in the urban planning process. Using Can Ricart as an example, we hope to provide elements for reflection on the logic of Barcelona's current planning process.

Poblenou: from the periphery to the centre

The history of Poblenou, a neighbourhood within the Barcelona district of Sant Martí, is closely linked to the industrialization of the Barcelonan Plain. Its position outside the walled perimeter of the ancient city, in a sparsely populated area with high levels of water in the subsoil, meant that after a long period of agricultural use, the municipality was most suited to heavy industry. During the first third of the nineteenth century, the area was developed largely for textile industries with factories and housing for their workers. Sant Martí was transformed into an industrial district with a heavily working-class identity, becoming known as the 'Catalan Manchester'.

The first major urban plan for Barcelona, the Cerdà Plan, involved breaking the city walls and connecting the ancient city in an orthogonal design with the five surrounding towns of Sants, Sarrià, Gràcia, Sant Andreu and Sant Martí. The successive reclassifications of land in Sant Martí (from farming to industrial use) consolidated its peripheral relationship to the rest of the city. This isolation led to a strong sense of belonging and identity among the inhabitants which crystallized in a dense network of largely anarcho-syndicalist associations and protest groups, characterized by their focus on the labour union as a potential force for a revolutionary change. In Catalonia, the Confederació Nacional del Treball (an anarcho-syndicalist trade union) hegemonized the left-wing movement until the end of the Civil War (1936–1939). These were ideal conditions for the implantation of a strong cooperative tradition that is still visible today.

The arrival of Fordism in Barcelona in the second half of the twentieth century precipitated sweeping changes to the locations of production, and an exodus of industrial activities towards industrial estates in the outskirts of Barcelona (Zona Franca and Baix Llobregat). As a result, between 1963 and 1990, Poblenou lost more than 1,300 industries, initiating a cycle of urban degradation and municipal neglect. Although the departure of industrial activities left the neighbourhood with large areas of abandoned industrial buildings, its continuing classification as industrial attracted neither property promoters, who preferred to build in the working-class housing estates on the outskirts of Barcelona, nor the attention of the municipality, which was more interested in the outward growth and renovation of more well-to-do neighbourhoods. Poblenou was relegated to the status of abandoned neighbourhood.

The presence of big empty areas and lower rents attracted new actors, who colonized the space and changed its traditional uses. In our opinion, three of these new uses are still important today: entertainment (a new night-time leisure zone of discos, pubs and live music bars has appeared), transport (logistics and transport companies took advantage of Poblenou's centrality and established bases here), and artistic-artisanal (workshops and studios for artistic production began to move into the neighbourhood from the early 1990s). The arrival of this last category in particular has contributed to the generation of a number of 'creative clusters', including within Can Ricart.

The 1992 Barcelona Olympic Games, and the remodelling of the city before and after the Games, changed Poblenou's situation once again. The creation of the Vila Olímpica and the regeneration of the seafront (Bonet-Martí 2004) were followed by new planning projects: the High Speed Train station in Sagrera, the Diagonal Mar project, the development of Glòries and the AGBAR tower as a new urban centre and icon for Barcelona and the Forum of Cultures. These now surround Poblenou, transforming its relationship to the city from peripheral to central, and making it highly attractive to property capital (see Figure 12.1).

Major pressure came from property promoters for the land to be reclassified for urban uses and housing. The municipal corporation, however, was more interested in the area's revitalization as a production district, believing it presented an opportunity to adapt to the new requirements of the 'knowledge society' and to improve Barcelona's position in the global economy (Trullen 2000). This thinking formed the basis of the '22@bcn' project, which aimed to exploit the new logic for the placement of Information and Communications Technology-related industries, and to compensate for the increasing tertiarization of the city by turning Poblenou into Barcelona's new technology district. Approved in 2000, the 22@bcn project materialized as a Modification to the Pla General Metropolità or General Metropolitan Plan. Its intent was to transform Poblenou's industrial area, previously called 22a, into '22@ zones', with the object of attracting knowledge-intensive industries. The plan also introduced a new class of facilities called 7@, which are publicly owned but privately managed. These facilities are dedicated to research, education and business related to new technologies.

In all, 22@bcn led to 1,982,700 square metres being made available for new production, public housing, green spaces, 7@ facilities and technical services

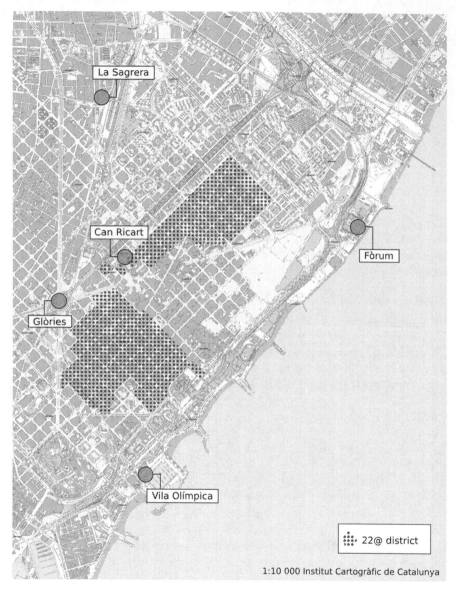

Figure 12.1 Map of Can Ricart, Barcelona.

Source: Map © Institut Cartogràfíc de Catalunya.

(Ajuntament de Barcelona 2000; Oliva 2003). The plan is under public control but the funding and development come from real estate investors.

Early implementation of 22@bcn required demolitions and displacement of the local population, leading to strong local criticism (Associació Veïns Poblenou

2001, 2002a, 2002b). Anti-capitalist groups claimed the plan would destroy the traditional composition of the neighbourhood for the benefit of large corporations and real estate agencies. Groups of residents who were directly affected wanted to be able to stay in their homes or at least be better compensated, and contested the changes. The local neighbourhood association took an intermediate position. While agreeing with the need for regeneration to improve local quality of life and maintain the productive character of the area, the association argued for respect for the architectural and historical heritage, and for the local population. This broad-based opposition to 22@bcn fragmented under many different agendas, but made a unified stand at Can Ricart.

Can Ricart, a thorn in the side of 22@bcn

Can Ricart is a neoclassical-style industrial site designed by the renowned academic architect Josep Oriol Bernadet between 1852 and 1854. It was completed ten years later by the architect Josep Fontseré i Mestres, the designer of Ciutadella Park and the Born Market. The factory was one of the first to be used for mechanical stamping in Catalonia, and one of the most important industries on the Barcelona plain. Although it was first used for the production of cotton stamping, from the 1920s on it became a rented industrial estate and was the site for other innovative industrial activities, such as the production of coconut oil by the Hispano Filipina company. Other industrial and artisanal activities were added later, and by the 1990s, it was home to a number of arts and cultural production studios, such as Hangar, an audiovisual company managed by the Associació d'Artistes Visuals de Catalunya and co-financed by the municipality.

Nowadays, Can Ricart occupies a space more than two city blocks (19,224 square metres) which makes it one of the biggest industrial sites in Poblenou and a key element of both Barcelona and Catalan industrial heritage, despite its rapid, recent deterioration. The process following the approval of 22@bcn can be summarized in the following four stages.

The 'tabula rasa' of the new planning process: the PERI of the Parc Central

On 29 October 2001, the third of a series of Plans Especials de Reforma Interior (PERI, or Special Interior Reform Plans) was passed for 22@bcn. It affected Parc Central, an area of more than 100,000 square metres in the north of Poblenou. The plan proposed the addition of an important centre of activity in front of the future Parc central del Poblenou (Ajuntament de Barcelona 2000:14). It is on this land that Can Ricart is located. The PERI proposed the retention of a number of unrelated elements of Can Ricart: the chimney (the only element in the city's heritage catalogue), the clock tower and two industrial buildings (see Figure 12.2). The rest of the site would be demolished and partitioned to improve the continuity of one street. The general characteristics of the plan were defined by the prestigious architect Ignasi Solà-Morales and approved without any political controversy or official complaints by citizens. At that time, the efforts of neigh-bours' associations were focused on another problematic PERI closer to the old centre of Poblenou.

Once the Parc Central PERI was passed the land was reclassified as part of 22@, allowing for a wide range of uses including office space. This promised major capital gains for the owner, the Marquis of Santa Isabel. A study was made of 22@bcn's financial feasibility, but no historical or architectonic study was made of the production activities that would be affected should the plan be imple-mented. No consideration was given, for example, to the fact that there were 34 companies located in Can Ricart at that time, employing a total of 250 workers. In addition to the companies, there was a bar/restaurant and two art studios (Can Font and Hangar). All these facilities were in rented spaces. Hangar was the only site considered for preservation because it was located in a rented space owned by the city council and had an agreement with the city council.

The defence of the activities and heritage of Can Ricart

The owner of Can Ricart began to arrange for increasingly shorter rental contracts in order to avoid the possibility of having to pay compensation to his tenants. When it became clear to the various companies that their contracts would not be renewed and that they would be forced to move out without compensation, they organized, forming the Associació de treballadors i empresaris de Can Ricart (Association of Workers and Employers of Can Ricart). In early 2005, they held a demonstration demanding fair compensation for all the companies and activities on the site (regardless of the types of contracts they had with the owner). This change of scale, from essentially a private conflict into a problem for the whole city, led to the more active involvement of over 40 cultural, educational, neigh-bourhood and youth organizations in Poblenou and across Barcelona. The Salvem Can Ricart (SCR) protest group was created (http://www.salvemcanricart.org), uniting demands for the conservation of the site for historic, identity and architec-tonic reasons with demands for the maintenance of the site's activities.

The more important members of the protest group included the Fòrum Ribera Besòs (http://www.forumriberabesos.net) and its heritage group, neighbourhood associations, architects, historians, the Association of Workers and Employers of Can Ricart and various youth and anti-capitalist organizations. Fòrum Ribera Besòs – a group of citizens and professionals dedicated to alternative planning models in the area between the Vila Olímpica and the Besòs River – played a fun-damental role in the process. Due to both the prior activism of its members and the production of reports and alternative proposals (Grup de patrimoni industrial del Fòrum Ribera Besòs 2006b), the Fòrum made SCR's demands more solid. In addition to the neighbourhood associations, a highly active and innovative group in SCR was formed by local artists based in the Can Font–Nau21 studio (http://www.nau21.nct/) and Hangar (http://www.hangar.org). These artists argued that Can Ricart should become an innovative urban space in the public domain.

One of SCR's first actions was to provide support for the threatened companies. Collective resistance prevented three attempts at forced eviction by the police. This served to publicize the issue throughout the city and diverse activities (guided visits, neighbourhood assemblies, international conferences, festive and cultural activities, demonstrations) made Can Ricart even more visible, in the process tightening the links to Poblenou's social network. Can Ricart became a symbol of creative neighbourhood resistance to speculative pressure and non-participatory planning.

The conflict prompted the Barcelona City Council, through its 22@bcn offices, to facilitate negotiations between the owner and the managers of the affected companies. The settlements were never made public but suspected increases in compensation and general legal pressure led to the companies gradually leaving the site. By the end of June 2006, the last company to have resisted eviction closed down for good as it was unable to cover the costs of relocation. Only one company managed to relocate within Poblenou; all others had to either move beyond the neighbourhood or close down (Grup d'Etnologia dels Espais Públics de l'Institut Català d'Antropologia 2006).

New plans and alternatives

During 2005 and 2006, SCR started fighting on new institutional fronts. Two important cases were the legislative initiative in the Catalan Parliament to preserve the heritage and historical memory of Poblenou, and a demand to the Generalitat's Department of Culture that Can Ricart be designated a Cultural Asset of National Interest (CANI). As a consequence of the former, in December 2005 the mayor cancelled all demolition licences at Can Ricart.

Meanwhile, SCR continued to explore alternatives. In-depth studies were made (Grup de Patrimoni Industrial del Fòrum Ribera Besòs 2006a) that sought an approach that would conserve the factory while respecting the rights of the owner. Suggested alternative uses included neighbourhood facilities (a selective waste disposal point and a kindergarten and community centre), city facilities (a labour museum and an education centre), social housing and an 'arts and creative centre' with more than 9,000 square metres of floor space dedicated to art studios and research centres. The resulting alternative plan included productive cultural and public uses of both local and global relevance, promoting Can Ricart as a node of urban centrality.

Little by little, the perseverance of the movement began to bear fruit. As a result of the mobilizations and meetings with local government officers and politicians, in April 2006 the council announced a new official plan for Can Ricart that was more respectful of its heritage and redefined possible uses. The new arrangements introduced by the plan were the following: the preservation of 60 per cent of the site, improved pedestrian usage, a community centre, widening the Hangar space, the location of the Language House (a project recycled from the 2004 Universal Forum of Cultures) and increased area for private 'lofts' (Ajuntament de Barcelona 2006a) (see Table 12.1). The new plan presented

Table 12.1 Facilities compared between the Can Ricart Plans

Salvem Can Ricart Proposal	City Council Proposal (April 2006)

A. Local facilities
    Neighbourhood centre + all-purpose
    hall + space for the Poblenou
    street committees: 2,100 m^2
    Kindergarten: 1,000 m^2
    Refuse collection facility: 64 m^2
	A. Local facilities
	    Neighbourhood centre: 910 m^2
	
	—
	—

B. National and municipal facilities
    Labour Museum: 4,000 m^2
    Live-in educational facility (CRAE):
    1,000 m^2
	B. National and municipal facilities
	    'Language House': 5,532 m^2
	—

C. Creative economy, arts centre
    Hangar Art Centre: 4,500 m^2
    Nau 21 Project: 1,275 m^2
    Trade and business incubator and
    cultural research centres: 3,225 m^2
	C. Artistic activity
	    Hangar Art Centre: 2,937 m^2
	—
	—

Total area of facilities (classification 7):
    17.164 m^2
	Total area for "@" facilities (classification
	    7@): 9,379 m^2

D. Housing
    Total subsidized housing area: 3,552 m^2
—
	D. Housing
	    Total subsidized housing area: 3,552 m^2
	    Unconventional housing in the heritage
	    precinct (classification 22@t,
	    LOFTS): 2,368 m^2

E. Economic activities
—
	E. Economic activities
	    Private production spaces inside the
	    heritage precinct: 5,765 m^2

Productive spaces outside the heritage
    precinct: 87,607 m^2
Total private use area: 87,607 m^2
	Productive spaces outside the heritage
	    precinct: 79,474 m^2
	Total private use area: 87,607 m^2

Source: Grup de patrimoni industrial del Fòrum Ribera Besòs (2006b).

advances. However, SCR members were concerned at the plan's failure to respect the unity of the space, the demolition of certain buildings, the insufficient number of creative spaces and social facilities, the addition of the Language House rather than the recommended Labour museum which was much more related to the site's history, and the proposed construction of a disproportionately large tower next to the complex. Nevertheless, this plan was approved in November 2006.

As a result of the Catalan Parliament initiative, in December 2006 a new Poblenou heritage plan was passed (Ajuntament de Barcelona 2006b). The elements to be protected (to various extents) increased from around 40 to more than 100. As for Can Ricart itself, the Heritage Plan declared the site a Cultural Asset of Local Interest (CALI) and consolidated the advances of the new plan announced by the council a few days before.

At this point, Can Ricart, with the exception of the Hangar, had been completely vacated. It was at that moment that an unexpected actor arrived on the scene: the Makabra group. This group specialized in circus arts and had been evicted just a few days earlier from a nearby factory. It squatted Can Ricart in December 2006 and avoided eviction by resisting a 34-hour police siege. After a few initial moments of tension and uncertainty, SCR decided to support the squatters as part of their protest against the loss of creative spaces in Poblenou. After eleven days, however, the squatters were forced out by the owner with support from the Council, one day before the Makabra group's activities were to be performed to the public for the first time.

The demolitions in a site declared a Cultural Asset of National Interest

A few months later, in March 2007, the Generalitat's Department of Culture initiated the declaration of Can Ricart as a CANI, classifying it as a collection of historical buildings. This is the greatest degree of protection that can be granted by the Catalan Government.

In the time leading up to the conferring of this protection, the site went through several changes of status. When the first council plan was opposed by citizen groups and action was initiated in parliament, the mayor declared a moratorium on all new demolition permits. The moratorium remained in effect until the CALI declaration and the April 2006 plan was approved. However, the initiation of the declaration of Can Ricart as a CANI meant that all new activity licences and demolition permits had to be approved by the Generalitat's cultural heritage commission. As a result, discussion reopened about those buildings not included in the CALI listing, and the possibility of a new plan for Can Ricart was raised again.

After several months of relative calm, in October 2007 demolitions started. Without SCR's knowledge, the Generalitat gave permission to private developers to demolish the buildings not included in the CALI, notwithstanding that these same buildings were included in the proposed CANI boundary. Urgent street actions and the initiation of administrative procedures to prevent the demolitions were to no avail. The only buildings that remained were those granted protection under the CALI and included in the council plan. The cohesive nature of Can Ricart was lost. The discussion now about Can Ricart has been reduced to the future of the remaining buildings and the use of the rest of the site.

The unequal impacts of mobilization

The case of Can Ricart, despite being only a small part of the city, had important city-wide impacts. In this final section, we will discuss three characteristics of the resistance movement: its organization, its collective action strategies and its impact on urban planning.

The organization and collective construction of knowledge through the heterogeneity of the movement was one of the strong points of the movement. Professionals and activists from many different disciplines dedicated time and ideas

Figure 12.2 Can Ricart tower, Barcelona.

Photograph by Marc Martí-Costa and Jordi Bonet-Martí.

within and outside the protest movement. The knowledge gleaned from urban planning, history, architecture, engineering and other sources combined to push the
protest movement's ideas forward and to guide new actions. Nine studies were made
of the site, with plans, maps, drawings and models generating extensive documentary support (see Grup de Patrimoni del Fòrum Ribera Besòs 2006b). The work of
SCR was recognized with three awards from various cultural institutions for its ideas
and perseverance in the defence of the history and heritage of Poblenou.

The combination of resistance through direct action and constructive viable
alternatives created a strong foundation when activists brought their claims
before the municipal and Catalan governments. Also, it went beyond the defence
of one specific site and provided new ideas about more democratic planning
processes, for example, through the focus on forms of urban centrality that are not
shopping centres or offices.

The debate over Can Ricart began as a localized struggle, but soon moved
beyond this to call into question the planning of the entire city. The issue was in
the foreground of social and political affairs, with the site becoming the focus of
heated public debate about the social, cultural and urban planning of the future
metropolis, and taking into account cultural heritage, social cohesion, the renovation of the city, and the productive interweaving of creativity and innovation.
It provided a catalyst for the convergence of many different neighbourhood, artistic and intellectual networks.

More concrete changes can also be seen. The changes can be summarized as successes and failures. The successes are several: the campaign around Can Ricart was able to preserve an important part of the site with the maximum protection possible. Moreover, the new buildings planned inside the site will respect the old distribution and proportions. There is an increase in the area reserved for facilities related to technological and social activities: the new approved plan creates a community centre and extends the Hangar space. The new Poblenou heritage plan gives more protection, and sensibility, to industrial heritage. A monitoring commission has been created, composed of members of local groups, professionals, members of cultural institutions and government officials. Finally, many ideas that were not put into effect at Can Ricart proved valuable at other sites. Very near to Can Ricart, the city council purchased two naves of L'Escocesa factory to maintain artists' activities. The soon-to-be-closed textile factory Fabra i Coats has been completely purchased by the city council and will be home to social and creative activities in the less central Sant Andreu district.

But the struggle for Can Ricart was not without its failures. The original activities and workers (before 22@bcn) were shut down or evicted. The plan did not include any help for them because they were considered obsolete, and because they had been working in a rented space. It is worth mentioning, however, that after the resistance, better compensations were conceded. The factory as a cohesive unit was broken, losing its historical sense and the opportunity to create an urban microcosm that could participate in the construction of a cultural and citizens' centre open to the public. The half of the facility near the park is maintained but the other is largely replaced by new lofts. All that remains of the former buildings is the odd important facade incorporated as a decorative element into the new private residences. The new plan limits the social and local facilities. The Labour Museum, the cultural project that was deemed by the neighbourhood association as a most appropriate use for this landmark of industrial modernity, was replaced by the 'House of Languages'. The final plan reduces the facilities for the arts and the creative economy, with the only creative industry remaining being the Hangar. The centrality and visibility of the original industrial buildings is reduced, in part due to the planned construction of large towers at the main entrances of the site.

Final thoughts

The Can Ricart conflict was the result of a top-down planning model. The process failed to consider the synergies that had developed between industrial and traditional craft activities and the new creative centres. It gave private developers a highly active role, in which their *tabula rasa* approach paid no or little heed to the social assets (production, heritage and intangible) of the territory.

In the case of Can Ricart and the 22@bcn plan, the focus was on transforming the activities and uses of the territory. Even though the majority of all new houses constructed as part of 22@bcn are social housing, the combination of the change in economic activities, the creation of new residential estates in the surrounding areas and the rise in the price of existing houses lead us to predict that the substitution of activities will be accompanied by a process of gentrification. Those new uses are part of the regeneration logic of the larger district with the perimeter of the Vila Olímpica, the seafront, the Forum area, the AVE station of Sagrera and the Glòries square.

In conclusion, we want to highlight three key aspects of this urban transformation. First of all, the council's plan can be considered an example of top-down logic attempting to artificially generate a mixture of uses, and ultimately producing higher opportunity costs than if genuine integration were sought between the practiced city and the planned one. Second, we hope we have shown that the emergence of a strong and diverse network of opposition, with creative practices and ability to propose alternative solutions, can, at least in part, modify municipal plans. However, in this case, real estate and property interests, along with the city council that depends on them for the development of 22@bcn, prevailed over the demands of the neighbourhood movement. Finally, analysis of the conflict over Can Ricart shows that defence of industrial heritage can be a major catalyst for urban mobilization. Nevertheless, if recognition of the value of heritage focuses only on the container, and the associated uses and ways of life are disregarded, the potential of the mobilization will always run the risk of being institutionally compromised.

Acknowledgements: We would like to thank Ricard Gomà, Abel Albet, Hannah Berry and Barbara Biglia for their comments and suggestions on a draft of this chapter.

13 The ambiguous renaissance of Rome

Giovanni Allegretti and Carlo Cellamare

Since 1993, with the establishment of directly elected mayors in city govern-ments, the Italian models of urban development have been deeply marked by the personal touch of their designers. Rome is no exception to this rule. Moreover, it is divided into 19 districts, called municipalities, with specific roles in some sector policies. Their citizen-elected presidents act as mini-mayors, adding their personalized visions of urban transformations to that of Rome's current Mayor Walter Veltroni, which some analysts have called 'Veltronism'.

Veltronism dreams of Rome on a grand scale as a global city, yet claims to be open to listen to the many voices of the Roman citizenry. *Modello Roma* (the Roman Model) is the name forged by the city administration itself to describe this particular style of government. This is marketed as being a strongly consensual approach to urban development that is striving for modernism, innovation and change (see AA.VV. 2007). The Roman Model, though, is essentially a model of continuous urban growth and development based on unsustainable land specula-tion and consumption which continually causes tensions between old and newly arrived inhabitants.

This chapter tells the story of a struggle against this growth-oriented model to produce alternative visions for the city and develop more participatory planning processes. We conclude that while this struggle has succeeded in small ways to counteract political shortcuts of the policy emphasis, it has as yet been unable to alter the basic premises of the prevailing urban development model.

The Roman Model

Modello Roma purports to recreate Rome as a 'competitive city of solidarity' (Allegretti 2008:3). Thus, the policy seeks to achieve two potentially conflicting aims: improving quality of life for city dwellers, and making the city more glob-ally competitive.

Policies to improve citizens' quality of life include the promotion of the multi-centric city (through preserving the diversity of neighbourhoods and the agricul-tural fields or parks which often lay between them); recovery of the degraded urban fringes and their public services, creation of employment 'incubators'; pro-tection of biodiversity and ecological corridors; conversion of redundant military facilities for civilian use; and the enhancement of minor archaeological areas, which are often protected but still not accessible (especially when located in the suburbs). The overarching 'participating periphery' project is part of Modello Roma's progressive social policies, aimed at increasing civic pride and participation through improvements to physical infrastructure in marginal areas and more widespread opportunities for social dialogue regarding design and development.

Policy objectives aimed at making the city more competitive include strengthening security, improving urban furnishings and lighting, rehabilitation of historical centres, increasing residential use in the inner city and attracting international commerce and tourism through grand architectural schemes and cultural events.

Between these two policy objectives, the latter undoubtedly prevail such that the whole urban and social fabric becomes secondary to the importance of landmark projects. This leaves a series of 'black holes' in Rome's renaissance where the most sensitive problems are left unaddressed. One particular 'black hole' is that of urban transport and mobility, where a general commitment to improve rail-based transportation has never seriously been put into practice. This is especially serious in a city where a mere 18.2 per cent of journeys are made on public transport and private vehicle use has boomed, resulting in an abnormal increase in individual mobility and commuting by workers forced to live farther from the city centre (Sartogo 2007:115). This represents how genuine commitment to improving inhabitants' quality of life contradicts the focus on landmark architectural schemes to make Rome globally competitive.

The Historical Centre as a symbol of urban struggles

As the capital's poster child, tasked with promoting the city's image internationally, the historical centre of Rome (almost entirely governed by the First Municipality district council) has been subject to major revitalization projects. City Hall asserts that this revitalization automatically brings positive benefits, indirectly to the entire city, but directly to the historical centre whose residents *should thus consider themselves privileged*, as implicitly stated in several public speeches by local government members. In reality, these policies benefit a small minority. Meanwhile, ordinary residents are faced with soaring real estate prices, evictions, the unravelling of the social fabric, increasing noise and air pollution, and shortage of resident parking. Massive increases in rental prices (today accounting for over 70 per cent of an average family's income, see Caudo 2007:98) has caused widespread displacement of traditional inhabitants. Further, many traditional artisan workshops have been replaced by hotels, bed and breakfasts, and fast-food chains to cater for the 23 million tourists coming through Rome annually since 2006.

Typical of these contradictions between tourism-centred policies and resident needs is the elimination of traffic in the Trident area to create the pedestrian squares termed *salotti di Roma* (open living rooms). This has caused widespread diffusion of cafe tables occupying public space, and residents complain of the loss of entire streets and squares such as Campo de' Fiori and Piazza Navona.

Such enhancements result in a de facto privatization of public space, generating heated debates among the remaining inhabitants. In Piazza Madonna de' Monti, residents are now obliged to ask permission of the cafes whose tables occupy the square to carry out their traditional neighbourhood festival.

Such clashes between the needs of local residents and urban policies have made the historical centre of Rome an interesting laboratory of urban conflict. The 'Transform!Italia' action/research group partially mapped these conflicts and their high degree of dynamism in a book and interactive Web site *La riva sinistra del Tevere* (Transform! Italia 2005).

Several historical *rioni* (neighbourhoods) on the fringe of the city centre – such as Monti, Testaccio, Celio and parts of Trastevere – continue to face gentrification pressures as they are attractive to well-to-do home buyers for their relatively intact social fabric, human scale of built form, visible local identity and proximity to the historical centre while offering distance from the suffocating crowds of tourists. In these neighbourhoods, there remains resistance to the voracious trends of urban transformation of formerly working-class areas, particularly through the action of some of Rome's squatted and independently run social centres.

The next section focuses on one of the most well-organized bottom-up initiatives which grew up at the fringes of the city centre. Its birthplace is in the Monti neighbourhood, yet it is highly representative of other struggles as it was able to get policy proposals approved for other surrounding areas in the city centre.

From the Laboratory for Urban Choices to the *Casa della Città*

In spring 2002, the Rome administration presented a new masterplan, including proposals for the whole town which had to be evaluated by the 19 district councils, as the Roman procedures provide. It was approved in 2006, after protracted public debate. This process of debate developed strong connections between a number of citizens groups and the First Municipality, whose young administration was open to new forms of public involvement in local decision-making. Thanks to the support of a group of university researchers who took on the role of guarantor, a Laboratory for Urban Choices was born. The new forum was open to involvement of individual citizens, as well as representatives of environmental associations, unions and citizen groups.

In the first phase, debate focused on the historical centre and resulted in the publication of several essays, such as 'Quality and Liveability in the Historical Centre' (Laboratorio sulle scelte urbanistiche nel I Municipio 2003). Then, a set of proposals addressing themes of commercial and residential quality were developed. The laboratory next confronted the problems of mobility, arts and crafts production, public space and pedestrianization, by mapping the problems and design proposals, and collaboratively making several critical observations of the masterplan.

In a second phase, discussion shifted to city-wide policies on commerce such as the salotti and also urban transportation. For much of 2005, the Laboratory was dedicated to the project Sbilanciamoci (Let's Un-Balance!). This experiment in participatory budgeting defined proposals for urban transportation interventions to insert in Rome's 2006 budget.

In May 2004, the Casa della Città (City Home) project was drafted, proposing to increase the stability and visibility of the Laboratory through the development of a permanent physical home. This was officially opened in April 2006 with support of the city council. Since then, the Laboratory disappeared and its activities were undertaken by the Casa della Città (www.casadellacittaroma1.org), which became the pivotal centre in the area for generating social dialogue and mediating between citizens' proposals and institutional policies.

In the meantime, the Laboratory's promoters were involved with the drafting of a set of rules for participation, aimed at allowing better participation by citizens in urban policy-making. They pushed successfully for the requirement that each municipality have a Casa della Città.

Today the Casa della Città has various functions. It makes information, documents and design proposals related to current urban transformation available to the public and provides qualified personnel to help local citizens understand this information. It also brings together the ideas and creativity of individuals and civic organizations through such competitions, training sessions and artistic events. Finally, as a public place devoted to promote and host debate, it provides logistical support for participatory processes related to policy-making in the First Municipality.

From the start, the Casa della Città organized itself into three working groups around the principal themes which emerged: neighbourhood markets and commerce, mobility, and urban revitalization. Recently a new experimental 'outreaching' methodology suggested to organize frequent meetings throughout the city (on specific themes) to encourage widespread citizen involvement. New life was given to the Sbilanciamoci project, including participatory design projects for revitalization of several streets.

Strategies and outputs

The Laboratory and the Casa della Città depend on the mobilization of grass-roots movements and the social fabric's capacity to stimulate public institutions to continuously re-discuss Rome's urban renaissance and its ambiguities. Only concrete outputs and impacts can engender enthusiasm for new struggles, and in this the mixed composition of the organizations is a pivotal strength.

Two short examples highlight this. They involve local struggles against some weak aspects of City Hall's renaissance strategies. One struggle concerned the commercial occupation of public space. When facing the invasive plague of cafe tables, the Laboratory proposed a serious evaluation of its negative effects, by documenting other Italian cases and critically reading the salotti di Roma regulation. Participants called for an extension of the 'plan for maximum occupation' (PMO) throughout the entire historical centre. This tool was established to regulate the space designated for cafe tables in relation to their distance to monuments, sight lines, and rights of way, but had previously been applied to only about 50 protected historical squares. The Laboratory called for the city council

Figure 13.1 Public demonstration in Piazza Madonna dei Monti, Rome.
Photograph by Carlo Cellamare.

to delegate to the municipality the development of new plans by way of participatory competitions.

Unable to get feedback from the city council, the Laboratory participants gained
the support of the municipality, which utilized a little-used procedural tool to force
the city council to provide a formal response to the citizens' proposal. By the end
of 2006, more than a year later, the PMO was approved and the city regulations
were amended (adopting almost all the recommendations of the Laboratory). In this
struggle, which involved coordinated action of the municipality and other civic
associations, the pivotal role of the Monti Social Network emerged. In June 2005,
this network opposed the increase in cafe tables in Piazza Madonna dei Monti,
which was partially illegal as the area used by bars and restaurants exceeded the
size authorized by municipal permits. The Monti Network invited all inhabitants to
occupy the square with their own tables and chairs, serving drinks to everybody
(see Figure 13.1). Positive participation of local residents was above all expectation. The event's success and its disturbing visibility coaxed the mayor himself to
come to the square to meet the inhabitants and formally accept a more restrictive
plan of maximum occupation, together with a tighter control of its implementation.

The second struggle concerned the lack of attention to urban transportation and
mobility problems. In this example, Monti Social Network again emerged as an

important catalyst even when divergent local interest groups such as artisans, merchants, taxi drivers, hotels, wholesalers, couriers were individually lobbying, and thus continuously undermining the unity and strength of the Laboratory proposals. Meetings between City Hall technicians and the Monti Social Network resulted in continuous indecision, so the mobility group of the network organized a series of community meetings to review the project. In addition, they hung sheets and banners between buildings and set up roadblocks. The protests were accompanied by a proposal to assign the recently formed Casa della Città to oversee a participatory design process.

The Laboratory work has thus come to fill an empty space of alternative proposals that should be offered by the public institutions. It is no coincidence that, despite maintaining critical distance, the First Municipality uses the Laboratory's presentation material every time it is called upon to express an opinion on city council decisions. This became especially pronounced after the Sbilanciamoci programme made concrete proposals for physical interventions of road maintenance, pedestrian zones and changes in circulation patterns in the central areas of Monti, Testaccio and Aventino.

The Laboratory's proposal managed to favour slow, low-impact transportation and pedestrian footpaths, limiting access to certain parts of the area to public transportation only (and then only electric). The scheme started in some unsafe streets (see Figure 13.2), where it was even difficult for residents to get out of their front doors, and it gradually took the shape of a policy to resist the devastating impact of traffic in terms of noise, air pollution, liveability and security. The best outputs were reached in the Monti neighbourhood – which had always been the very active centre of all struggles and of some self-managed experiments for implementing new measures – but their impacts slowly broadened to the surrounding neighbourhoods within the First Municipality.

In fact, at the end of 2006, a decision by the municipality (valorizing the Casa della Città's proposal) and the threat of further roadblocks by the Monti Mobility Group finally caught the attention of the mayor. Even the most innovative of the proposals have now been approved, including the extension of the daytime and night-time limited traffic zone and the creation of low-speed traffic zones.

These partially successful struggles clearly pointed out that a tension exists within the wide range of institutional strategies for coping with city problems. On one side, in fact, City Hall policies – guided by Mayor Veltroni's vision – regards the historic centre as the very *locomotive* of Rome's economic developments, having investors and tourists as a main target, while underestimating the importance of its inhabitants' daily quality of life. The powerful arm of this strategy is the Office for Historical Centre, a special structure which was created to coordinate planning and management policies in the heart of the Eternal City. On the other side there is the under-resourced First Municipality, which is directly elected by the historical centre inhabitants and owes to them accountable policies aimed at bettering the daily liveability of their territory, at least in the thematic fields which were devolved to districts' control, such as public space schemes, local mobility, culture, social assistance and green areas. The permanent 'tension' provided by

Figure 13.2 Protest against traffic in Rione Monti, Rome.

Photograph by Carlo Cellamare.

these two complementary (and sometime antithetic) visions sometimes results in a monolithic block which defends the interests of powerful economic lobbies; in other cases, it opens conflicts between the stronger and the weaker of these public institutions. And the latter requires tactical alliances with the social fabric, especially to strengthen the implementation of resident-oriented policies.

Another clear issue pointed out by the outcomes of the Laboratory struggles is that the mix of different actors is a strong resource for all the initiatives which want to have a say on public policies. In fact, on some occasions the bottom-up mobilization of inhabitants stood as a key factor for catching the attention of the mayor on district problems. Further, the First Municipality was sometimes able to find institutional ways of defending residents' needs within City Hall decisions.

Between conflict and counter-culture: some broader impacts

The First Municipality's Laboratory on Urban Choices is a space of confrontation which, while not attracting the general public, often brings together the representatives of local associations which act as conduits to wider social networks and residents' committees. It is an umbrella group whose added value lies in the articulation of protests and alternative proposals, stimulating the municipality and the city administration to constantly strive for more innovative policies which could democratize the Roman Model.

Beyond these concrete outputs, other impacts have been felt. The Casa della Città, as with the other 17 Territorial Laboratories which have been active in Rome in the past seven years, plays a role in cultural and political production, to continuously reproduce counter-cultures that help nurture a different way of thinking of the city. While the city administration has still not formally recognized its existence, for the municipality (which acts as institutional moderator) the Casa della Città is an indispensable foundry for ideas, defended fiercely by its inhabitants as the principal participatory space at the municipal level. It sustains opposition to the dominant urban renaissance policies, acting as a critical consciousness both through its own activities and its role as an incubator of other initiatives, projects and campaigns. Finally, it carries out important functions of connection and coordination between various associations, committees and social networks that bring vitality to the historic centre. Connecting differing realities into a network helps to 'rebalance' civil society: stronger associations, who could carry out lobbying autonomously, found themselves confronted with differing views, while smaller, weaker ones found support in the wider, structured network of action. In some cases, the Casa della Città helped activate new associative networks and provided overarching coordination for initiatives.

Today, the Casa della Città has no formal legitimacy, nor are its projects enforceable or obligatory, though it derives some authority from the rules on participation to which many of its participants contributed. Some transformations in the institutional frameworks of government are progressively strengthening these alternative 'institutions of mediation' but their strength mainly relies on their creativity and coordination of different grass-roots actors. There are, of course, always new battles to be fought, but the presence of the Casa della Città provides a stable conduit between local people and the institutional frameworks of government.

Windows of hope

These stories of struggle have demonstrated the ability to react against an urban renaissance guilty of marketing only its positive attributes, without allowing a debate concerning its contradictions and side effects to flourish. It is worth underlining the sense of hope produced by even small experiments such as those described here for the First Municipality.

What is more, the phenomenon has grown beyond the boundaries of the historical centre, influencing residents of other municipalities and representing an example for others to turn to for inspiration. The clear message is that if such objectives could be achieved in the historic centre of Rome (the hardest hit by the soporific effects of the urban renaissance), there are even greater hopes for other zones, especially those in the fringes of the city, characterized by a decades-long history of vitality and activism.

The Roman Model, even though it is open to local feedback, poses a strong inertia against transformation of its structural tendencies, especially when this is articulated through conflict and counter-proposals. That is why it is important that in January 2008, more than 20 Roman groups – representing dozens of grass-roots organizations engaged in initiatives similar to the First Municipality Laboratory – gathered in the umbrella committee 'Freedom and Participation' to give more critical weight to their claims for the respect of the 'procedural' rights of citizens to participate in urban policy decision-making.

These rights appear to be formally recognized and marketed as the distinguishing core of the Roman Model and some of them are formalized in the 'Rules of Participation' document, approved in 2006. Yet the reality is different, and participatory processes often appear little more than lip service beneath a predefined, top-down urban strategy. Even the recent multiplication of participatory processes promoted by City Hall is not a guarantee, and we argue that a strategy of numerous small and site-specific moments of consultation simply divert social debate away from wider and more important strategic policies (Allegretti 2006; Cellamare 2007).

The level of conflict with the city administration must always be kept high and procedural victories that have been won should be considered no more than a start. Proof of this lies in some of most progressive resolutions approved by the City of Rome since 2004, such as those on ethical sponsorship and on creation of the 'City of Alternative Economy'. The first committed City Hall to carefully select private sponsors for municipal initiatives, while the second provided funding for the creation of a public-run complex of buildings which act as an 'incubator' for cooperatives and firms working in the fields of organic agriculture, fair-trade and responsible tourism.

In each of these resolutions, the city responded to a specific grass-roots campaign, taking on a courageous commitment to innovate its own policies in order to promote morally responsible behaviour, to protect the interests of the less privileged and to give to 'borderline issues' a broader visibility in the public space. However, to achieve implementation of even the least of the objectives contained in these commitments has required continuous debate, action and struggle. For example, the consortium of grass-roots organizations which runs the City of Alternative Economy has continuously to fight against the attempts of City Hall to transform it in a mainstreamed wealth-production enterprise, while *Osservatorio*

Oppidum (a special bottom-up-run structure) had to be created in order to monitor and critically assess the widespread disrespect of municipal commitments in the field of selecting sponsorships.

An open conclusion

Within such a framework, the experience of the Casa della Città undoubtedly points out some risks and challenges of coping with the Roman Model. The model is both a stiff growth-oriented strategy (particularly when dealing with the historic centre *resource*) and a platform open to making minor adjustments in order to better balance its two main strategies: improving quality of life for city dwellers and making the city more globally attractive and competitive. The latter includes the lowering of urban tensions (which are seen as scaring investors) and thus suggests the implementation of tools and spaces for listening to the citizens. These are often used to *conquer* bottom-up struggles coming from a very active social fabric, which has then to fight to enable them to survive as places for the production of alternative visions for the city, as opposed to small compensative actions to reduce the negative impacts of the mainstream model.

From this perspective, the Casa della Città experience shows that the most difficult task is maintaining unity among the different actors of the social network, in order to transform tensions and different perspectives into a single voice. This is necessary in the face of City Hall's '*divide et impera*' strategy, which is to consult historic centre residents' groups in individual meetings in order to break the unified front of residents. It also shows that modifying some excesses of the Roman Model is possible, but only via a permanent redefinition of creative struggles and a variable geometry of tactical alliances with economic and institutional stakeholders (such as the First Municipality district government).

In this respect, the experience shows its limits. In fact, while a permanent mobilization of citizens is difficult to achieve, translating temporary tactical alliances with organized stakeholders or district institutions into permanent strategic ones seems impossible, due to conflicts of interests between them and the residents.

The conclusion of such a tale is probably that ameliorating some of the worst excesses of the mainstream policy emphasis is possible. Sensitive problems left unaddressed by general city planning, such as traffic, were partially solved, and ambiguities such as the overexploitation of public spaces for commercial purposes were tackled creatively. But the basic premise of the prevailing urban development scheme remains difficult to alter, even after such determined urban struggles.

14 Struggling against renaissance in Birmingham's Eastside

Libby Porter

Urban policy-makers in the United Kingdom are currently enthralled by the urban renaissance agenda. Coupled with the very wide-ranging powers for compulsory purchase (or eminent domain) and a lack of third-party appeal rights, this makes urban renaissance policies very powerful place shapers indeed. However, making those policy agendas work in practice requires that policy-makers re-imagine the places in their jurisdiction as essentially empty. Most places are, of course, not 'empty' at all but full of all kinds of different life: local residents, local businesses, a range of human activities, not to mention the habitation and use by non-human residents in old industrial areas rich with biodiversity.

This chapter details the peculiar story of urban renaissance as it is unfolding in Birmingham, the United Kingdom's second biggest city. In this story, a large tract of brownfield land is being assembled for presentation to the market for development as a new sustainable, cultural and knowledge economy quarter for Birmingham. In the way stand one long-time resident and a handful of established businesses, all of whom would like to participate in the future of 'Eastside', and all of whom objected to Birmingham City Council's compulsory purchase order on their property. The story highlights the power of assumptions underpinning contemporary urban renaissance policy, as well as the possibilities and politics of struggle for those who stand to lose the most from urban renaissance agendas.

As a regeneration area, Eastside is a 130-hectare section of Birmingham lying to the eastern edge of Birmingham's city centre. In actuality, Eastside is comprised of two inner city districts (Digbeth and Deritend), and the industrial birthplace of Birmingham (some claim of England). As an old industrial quarter of the city, it has a gritty urban feel and is criss-crossed by huge bluestone railway viaducts and a canal and lock system, both so important to Birmingham's industrial dominance. The district is made up of a unique blend of old factories, heritage buildings, dilapidated warehouses, old terrace and canal housing, corner pubs and chip shops. Since the 1970s and the housing policies that forced the relocation of thousands of local residents to peripheral housing estates, both Digbeth and Deritend have steadily lost population and are now home to only a few hundred people.

Creating Eastside as a blank slate

It was not until the mid to late 1990s that the district became the centre of Birmingham City Council's regeneration attention. Following the 'success' of major physical improvements and new developments on the western edge of the city centre, the council had developed a series of plans for the regeneration of different 'quarters' around the city centre. Then known as the 'Digbeth Millennium Quarter', Eastside's new future began with plans for the 'reintroduction of street frontages, mending the urban scar created by the former Inner Ring Road, neighbourly use of scale and mass-ing, establishment of landmarks, re-enforcing character, harnessing canal potential and the creation of legible routes' (Birmingham City Council 1996:3). By the late 1990s, the area had been re-branded Eastside and the drive for regeneration began to be felt. After a successful bid by the council and partners for funding from the European Regional Development Fund (ERDF), the idea of *sustainable* urban devel-opment became a brand for the regenerative mission in Eastside, generally cast in environmental terms such as the adoption of built form technology such as green roofs and combined heat and power schemes (see Porter and Hunt 2005).

Regeneration initiatives are predicated upon the characterization of an area as needing renewal – they are a 'problem' requiring state intervention. Eastside was no exception. Birmingham City Council's assessment of Eastside in an early strategic framework described Eastside as having

> an essentially industrial character. Small scale engineering and metal work-ing with some warehousing dominates. The southern part of Digbeth is home to many long established firms several of which occupy older prem-ises that are no longer ideal for modern activities. A number of firms also operate from restricted sites. Dereliction is evident in places ... the general environment is cramped and congested.
>
> (Birmingham City Council 1996:5)

A map accompanies these statements in the plan, showing gateway opportunities, industrial heritage areas, open space opportunities and major underutilized sites (Birmingham City Council 1996:6). In the updated regeneration plan, known as the Eastside Development Framework (EDF – the key planning document to guide development), similar techniques are used. Indeed, Eastside is described by (then) Mayor Albert Bore's foreword as a 'phoenix' (Birmingham City Council 2001), suggesting something new arising out of the ashes of dereliction and decay. Throughout this document, the area is portrayed as 'tired', with 'empty and under-utilised buildings', 'vacant sites' and 'poor quality developments'. Existing condi-tions are championed only when they relate to the industrial or architectural heritage of particular parts of Eastside, or when they relate to recent developments such as Millennium Point and the emergence of a high-technology and learning precinct proximate to Aston University. For urban policy-makers, then, Eastside is tired, run-down, dirty, lacking in economic activity, vacant and derelict.

This discourse of decline and dereliction underpins the 'desire to plan' and provide a more orderly urban environment. Urban dereliction is indeed often a problem, although it offers some benefits, as recent research on the unexpected benefits of derelict sites for urban biodiversity (The Wildlife Trust for Birmingham and the Black Country and University of Birmingham 2005) and marginal cultural activities (Shaw 2005b) has shown. Thus, Eastside is portrayed in the primary planning frameworks as essentially 'empty', with very little in the way of use, activity and meaning other than some poorly utilized industrial sites and a strong architectural heritage. Eastside is rendered a 'blank slate', ripe for development because of its 'underutilization' and proximity to the city centre, and 'easy' for development because of the apparent lack of any existing uses and conditions standing in the way of development. Imagining Eastside in this way is what makes it possible to draw the kinds of overall schematic land-use plans that are the bread and butter of urban regeneration and spatial strategic planning. As this place is rendered 'empty', it becomes literally possible to redraw it, to re-imagine the place as a different place in terms of use, connectivity and movement. Existing but silenced memories, desires, connections and activities are buried and forgotten.

The strategic regeneration framework for Eastside entirely ignores what is in fact an area rich in people's lived experience, both historical and current. There is an existing residential population, albeit small, who live in two small clusters of houses and those who live on the canal in longboats. There are numerous social networks and activities that centre around the old working men's pubs and other cultural venues. Yet none of these people or their activities and memories are mentioned in the planning instruments. The early strategic plan stated that 'there is very little housing in the area' (Birmingham City Council 1996:6), but by the time the later EDF plan was published, the local residents are not mentioned at all.

The EDF is not the whole story of Eastside, as any plan is rarely the whole story of any place. There are other Eastside stories that the plans do not show us, important stories of connection, meaning and everyday use. During research undertaken in 2006 with Eastside residents, it became apparent that the neighbourhoods that make up the Eastside area may indeed be run-down and gritty, yet people have a strong attachment to place and its history. Many residents have lived there for decades, raised their families there and expressed a feeling of being 'at home' in Eastside (though they did not call the neighbourhood by this name). People said they felt a strong sense of attachment to the place; as one woman said, 'I raised my family in this house'. Another resident described the importance of the canal system being near his back door as a place for leisure: 'my kids like to feed the ducks there'.

This attachment to place was mobilized when one long-time resident (the only owner-occupier in Eastside) received a compulsory purchase order from the council. In this story, we see how the myopia that is intrinsic to how the local state in Birmingham 'sees' Eastside, can result in pain and injustice, but can also be the catalyst for grass-roots struggle.

Mr Grove and the compulsory purchase order

Mr Fred Grove has lived in his canal-side house on Belmont Row (see Figure 14.1) in the Eastside area for over 40 years. It is the home in which he and his late

Figure 14.1 Home of Fred Grove, inner city Birmingham.

Photograph by Libby Porter.

wife raised their seven children. In the urban landscape of contemporary Birmingham, it strikes the casual observer as an unusual place to live, as it is by no means a residential street. The house is the only dwelling on the street and is surrounded by industrial and warehousing uses – a fairly noisy and unpleasant urban environment to the outside eye.

Mr Grove's home was one of many along the canal side – built by the water authority over 100 years ago to house canal lock-keepers and their families. He rented the house from British Waterways and then purchased it in 1992 when the canal authority put them on the market. Prior to living in his current home, Mr Grove and his family had lived in the next door property for 15 years. Up until the 1950s, around 20,000 residents lived in the area now described as Eastside in the new regeneration scheme – this includes the neighbourhoods of Digbeth and Deritend, and parts of Duddeston. By 1971, the population had fallen to around 550 people as a result of massive demolition of housing designated for 'renewal' in the post-war period under the Blitz and Blight Act 1944. The residual population declined further, by about 85 per cent to 2001, when the census showed a population of only 87 people. The post-1970s decline was a result of the deindustrialization of the Birmingham economy and the 'best first' policy of Birmingham City Council to re-house residents in 'renewal' areas, or peripheral council housing estates. Those who could not show an ability to maintain a property were left in the small numbers of social housing in the declining inner city neighbourhoods.

In the time that the Grove family called Belmont Row home, then, the urban landscape around them changed from a residential working-class neighbourhood supported by local jobs in nearby industries to a place abandoned by its population and rapidly deindustrializing. Since the adoption of the EDF, Mr Grove has found himself in the middle of a zone marked for considerable change and redevelopment, led by the local state. The council and the Regional Development Agency (Advantage West Midlands or AWM) began negotiations with him to purchase his property, but he was adamant from the outset he did not want to move.

In April 2006, a compulsory purchase order was made by Birmingham City Council (on behalf of AWM) to acquire those properties that remained in private ownership and were in the area marked for the Technology Park, City Park and Learning and Leisure Quarter development zones. The order included Mr Grove's property and also a number of other small businesses that had been operating for many years in the area – the Moby Dick pub, Rosa's Café and Los Canarios restaurant. All were well-established local institutions and the owners of each business had been expressing for many months a desire to be considered part of Eastside's future, rather than of its past.

The compulsory purchase order stated that the inclusion of locally listed buildings in the order, such as Mr Grove's home and the Moby Dick pub, was necessary to

> help ensure the area is redeveloped as a whole within an appropriate timescale and with both buildings integrated within the final development. Their retention within the development is likely to necessitate accommodation and other refurbishment works to ensure that they sit appropriately within the redeveloped area.
>
> (Birmingham City Council 2006: paragraph 3.13)

The council and AWM expressed no intention to demolish Mr Grove's home, despite wishing to purchase it from him. Indeed, quite the opposite. The house itself is of local architectural interest and was planned for retention as part of a 'thriving waterside community around the canal' (Birmingham City Council 2006, paragraph 2.8). Thus, it seems that it was Mr Grove himself being removed (not his property) as it is *he* that was considered to be ill-fitting in the new imagination of this part of Birmingham.

The primary 'reason' for the Compulsory Purchase Order (CPO) of Mr Grove's home is to 'help ensure the area is redeveloped as a whole' (Birmingham City Council 2006, paragraph 3.13). The language expresses an intent here to treat the 'zone' in which Mr Grove's home has fallen into as a uniform whole – as if there were no existing (or planned) variation across the neighbourhood. Yet in other sections of the Council's Statement of Reasons, it carefully claims to want to develop a 'sustainable mixed use development' (ibid.: paragraph 2.8) incorporating smaller scale 'character-development' around the canals and in the

Warwick Bar Conservation area, in which Mr Grove's home is located (Birmingham City Council 2002:7). In other policy documents, the varied and interesting nature of the canal-side urban form in Birmingham is noted for its uniqueness and conservation significance.

Yet, as the order states, it is more convenient if all the property is in council ownership (ready to sell on) in order to 'give certainty', 'enable the area to be developed in a comprehensive manner' and 'ensure the ability to gain vacant possession of all the land, property and interests' (Birmingham City Council 2006: paragraph 4). Comprehensiveness and certainty are key themes here. It appears that proper, efficient development cannot take place in a more piecemeal fashion, nor with existing interests remaining in place. In other words, to attract investor/developer interest, the council must create a 'blank slate' to offer to the market – a place with no memory, no people, no community, no spirit. Just a flat, uniform, empty patch of dirt, with all traces of earlier life expunged (except where they might usefully turn a profit).

One final aspect of Birmingham City Council's Statement of Reasons worthy of comment is the desire for any relevant existing forms to be able to 'sit appropriately' within the new development. This is a strategy that has been used before in Eastside with equally traumatic effects. Long-standing pubs were compulsorily purchased to make for the new City Park Gate mixed-use scheme in the first stages of the regeneration project (see Porter and Barber 2006 for the story). This suggests that the council has a clear vision for what the area will become in order to be able to assess what 'sits appropriately' and what will not. Such an approach suggests that the urban renaissance agenda for Birmingham is one of state-led gentrification, where the state seeks to remove unwanted people, activities and networks in order to 'tame' the inner city for the return of middle-class values, tastes and investment (Smith 1996).

Struggling against renaissance

Mr Grove and the small business owners that eventually became the subject of the compulsory purchase order had been quietly campaigning as individuals, directly to the council, against their removal for many months. Meantime, the council and AWM's thinly disguised gentrification desires had begun to raise concerned eyebrows in other local circles. It was at this point that the Eastside Sustainability Advisory Group (ESAG) became more publicly active on the issue of displacement. ESAG is an affiliation of different organizations incorporating local NGOs, government agencies, activists, researchers and built environment professionals. The group was established as part of the original successful bid the council made for ERDF funding, which required sustainable development to become a central feature of the Eastside regeneration programme. ESAG's remit was to advise the council on sustainability issues in the regeneration of Eastside. Birmingham City Council effectively sponsored the group providing space for meetings and, in the early days, assistance with administration. I was an active member of ESAG between 2004 and 2006.

ESAG had been concerned at the lack of apparent public consultation on plans for Eastside, and the extent to which the assumption of Eastside as devoid of residents and activity had become an excuse for inactivity in this area. The group organized two events to test the waters of civic interest in the issues during 2005. These included, first, a 'Grumble Sale' held on a warm June afternoon in an area of open space where Eastside meets the city centre. Participants were asked to share their stories and perceptions of places in Eastside, and identify where they felt public policy should be directed. A second event, 'Window on Eastside', was held later in the year, where the public was invited into the council offices housing the Eastside regeneration team to hear more about the proposals. Both events were organized by ESAG, with the support of Birmingham City Council's Eastside planning team.

Many issues were raised by the people who participated in these events, and it was here that the group began to hear the stories of small property owners in the area such as Mr Grove and the proprietors of Rosa's Café and the Los Canarios restaurant. It became very obvious to ESAG that there was considerable local concern and objection to the manner in which the regeneration strategy was unfolding, and frustration at the lack of opportunity for participation, or even basic consultation, in decision-making processes. ESAG advertised a public meeting in 2005 for local residents, workers and business owners in Eastside to discuss whether there was interest in forming a local group to more effectively represent local people's perspectives. Over 50 people turned up to the meeting, and together enthusiastically decided to establish a local action group, now known as the Eastside Community Group (ECG), which continues to meet and campaign.

Mr Grove attended this initial meeting and presented to the meeting the letter from the council and AWM he had then just received setting out the compulsory purchase of his home. The ECG decided to mount a campaign to assist Mr Grove as well as the other business owners who did not wish to be removed, using local media contacts. There was significant media interest particularly in Mr Grove's story (the 'old man being removed from his home' angle was an easy media winner) and through a savvy use of the media, very considerable pressure was brought to bear on Birmingham City Council and AWM.

Many of the statutory landowners also made formal objections to the compulsory purchase orders, including Mr Grove, and both ECG and ESAG also made objections as non-statutory parties. Due to the objections, a public inquiry was held in February 2007 to decide the matter. Days before the public inquiry opened, Birmingham City Council and AWM agreed to withdraw Mr Grove's property from the compulsory purchase order, and he has been allowed to remain the owner of his property. The inquiry went ahead for the other landowners to present their case, but found in favour of the council and approved the compulsory purchase of the remaining lands included in the order. The reasons for the council withdrawing Mr Grove's property from the CPO have never been made publicly available, though local campaigners consider the media interest in the case was significant.

Conclusion

Mr Grove's story exposes the ways in which regeneration activity generates a particular way of thinking and 'seeing' place that is profoundly myopic. Neither Mr Grove nor any of the other people whose livelihoods are built around Eastside have ever featured in the 'renaissance' discourse for this large regeneration area, and it was only when a grass-roots struggle objecting to the process emerged that any recognition was given to existing residents and businesses. Prior to this, those people and activities had been rendered simply invisible. This silence about existing life in a place is the moment that renders contemporary regeneration activity possible, as this activity is widely predicated on the assumption that regeneration is all about 'bringing back poor urban areas into productive use'. Local planning frameworks create blank slates, or empty places, to allow a rationalization of state intervention in these areas. Such myopia produces profound social and economic injustice. In this case, neither Mr Grove nor any other local people were considered worthy of renewal, and technical-legal processes such as CPOs have been applied in ways that sought to actively exclude those people from participating in the acclaimed benefits of urban renaissance.

The struggle against the processes of site assembly brought some success, especially for Mr Grove, who won his fight to remain in his home. This was a qualified success, however, in a wider context, as the compulsory purchase of other properties, all of whom could have remained viable businesses in the new Eastside, is now complete. Mr Grove remains in his home, surrounded by an increasingly vacant landscape as businesses start to relocate, or are closed down. Very soon, major construction work will begin to take place on his doorstep, though he is determined to remain throughout.

The success of his, and his supporters', struggle resided, first, in the existence of a local network of people who together had the capacity to generate local debate and action and, second, in the statutory powers available to landowners to object. Other residents still in Eastside who are all social renters and other 'unpropertied' people will likely be evicted much more easily with far less recourse for action.

It is unclear exactly why and how Birmingham City Council came to the decision to withdraw their compulsory purchase order from Mr Grove's property. What *is* clear is the underlying gentrifying desires of the urban renaissance agenda in Birmingham. While committed local people can successfully work against those tendencies, as has been shown in this story of the campaign for Mr Grove, it is fair to say that the gentrifying agenda remains intact. The gentrification, rather than the regeneration, of Eastside has only just begun.

15 Urban renaissance and resistance in Toronto

Ute Lehrer

Active 18 revels in the cultural, social, and economic diversity of the existing neighbourhood. But we are in danger of losing what makes the area special. Gentrification has become vicious. Three galleries have closed in the last few months along the stretch between Dufferin and Dovercourt (one replaced by a Starbucks) due to skyrocketing rents. Cultural workers are being priced out of the rental market. The tragedy is that the ultimate losers will be the people who move into the overly dense condominiums we are fighting. They are attracted to this neighbourhood by its current vibrancy. That vibrancy will be gone by the time they arrive.

(Active 18 2006:5)

Only ten years ago no one would have imagined that downtown Toronto would become a major hub of building activities. Today construction sites and cranes are indications of renewed interest in the urban centre. High-rise condominium towers transform the landscape, world-renowned architects such as Daniel Liebeskind, Frank Gehry and Will Alsop leave their marks on the extension of cultural and educational facilities, and the neglected waterfront is promised rejuvenation through large-scale mixed-use redevelopment (Lehrer and Laidley 2006). These attempts at reinvestment are supported by an array of policy documents that bring together the rhetoric of urban renaissance (City of Toronto 2003) with the concept of Richard Florida's 'creative class' (Florida 2002) and combine them with questions of intensification and sustainability of urbanized areas (City of Toronto 2002).

Toronto's most recent planning practice needs to be seen in this context. The story told here is the usual one: a city that sees itself on the same level as other major cities throughout the world and wants to compete to attract investment and highly skilled workers (Kipfer and Keil 2002); a city that has spent the past 20 years chronically under-funded, where manufacturing industries have been squeezed out, gradually departing for other places or shutting down altogether and leaving behind production sites for new users and usages. 'The city that works' as Toronto was called in the 1970s (Donald 2002) is now trying to reinvent itself by making a case for the creative industries in general and the arts and architecture in particular. Where better can this be achieved than through the renaissance of entire neighbourhoods?

This chapter looks at one particular case, the West Queen West Triangle (WQWT), and at how one part of the neighbourhood was galvanized to oppose the kind of planning currently seen in Toronto, where the planning department is overrun with proposals for high-density condominium development. It is a case study of a large-scale redevelopment project in a neighbourhood in Toronto that has become home to people working in the so-called creative industries, and thanks to the mix of a certain demographic and large stretches of underused industrial lands, is exposed to major pressures by the building industry. This case demonstrates not only how the typical pioneers of gentrification get squeezed out together with their working-class neighbours, but provides an example of how the pioneers, well aware of their roles as gentrifiers, are fighting back with a strong voice that is asking for 'good' planning, based on principles of socio-economic diversity, accessible public spaces and environmental sustainability, and promoting their approach with the slogan 'yes-in-my-backyard' (YIMBY).

Parkdale: birth, decline, neglect and revival

The redevelopment of WQWT needs to be discussed in the context of the Parkdale neighbourhood, an area that has experienced the familiar story of birth, decline, neglect and revival (Lehrer 2006). The neighbourhood is located southwest of the downtown core and is connected to it by a major streetcar route. Its roots go back to the nineteenth century when, driven by lower taxes and less crowded conditions than in many parts of Toronto, relatively affluent people settled in Toronto's first commuter neighbourhood and built Victorian and Edwardian style houses on large lots (Filey 1996). Incorporated in 1878 as the Village of Parkdale, it was annexed by the City of Toronto already in 1889 (City of Toronto 1976).

While the Great Depression left its imprint on the neighbourhood, it was particularly the construction of the Gardiner Expressway in the 1950s, an inner city freeway, which cut the neighbourhood off from its privileged access to the waterfront, followed by a building frenzy where a large portion of the housing stock fell victim to high-rise redevelopment that caused a downward spiral. Real estate prices dropped, speculative investment followed in the form of absentee landlords and investment firms, and larger homes were converted into rooming houses. The deinstitutionalization of more than 1,000 psychiatric patients from the Queen Street Centre for Addiction and Mental Health in the early 1980s added to the already strong presence of single households in the neighbourhood. In the 1990s, the media also did its fair share by publishing news reports, stigmatizing this neighbourhood as one of being home to crack houses, prostitutes and mental patients in a 'little ghetto of misery' where 'children are afraid to play outside' (Slater 2004a:313). This stigmatization helped to keep property values low for a considerable time.

This neighbourhood is unique in many ways: it has more than twice as many buildings erected before 1946 as other parts of the city and three times more tenants than owners, whereas city-wide the ratio is about 50:50. It is home to many low-income residents, single households, as well as an economically vulnerable but socially creative arts community. However, over the past few years, more and more affluent professionals have moved into the neighbourhood, led by the kind of urban setting that attracts the 'creative class', relying on 'place' to foster innovation and economic development. It appears that it is only a question of time until Parkdale as a socially mixed neighbourhood with predominantly low-income households will fall under the pressure of gentrification. While we have seen pockets of gentrification since the 1990s, overall Parkdale has been able to resist large-scale gentrification until very recently (Lehrer and Winkler 2006; Slater 2004a) with one exception: Queen Street West.

Gentrification pressures from the downtown core have moved along this major artery, converting former factories into much sought-after loft spaces on their way west, and now have arrived in Parkdale (Levitt and Adams 2005). Attracted by affordable spaces and the uniqueness of the neighbourhood, artists and other pioneers of gentrification have moved into the neighbourhood, taken up residence in former industrial buildings, opened store-front gallery spaces and upgraded some of the places. The buzz that came with these activities changed the neighbourhood into a happening place for culture and counter-culture, and ever since, the old greasy spoons and used appliance stores have been competing for their survival with stylish bars, galleries and boutiques. The neighbourhood is now known as having one of the largest concentrations of artists in all of Canada, and in early 2006, it received city-sponsored street posts, naming it the 'art and design district'. Attracted by its specific urban lifestyle, certain sections of the middle-class started to flock to the neighbourhood, a phenomenon that is well described in the local media: 'In any big city, art districts will come and go, charged into being by the fresh infusions of energy and inspiration that only artists bring, only to be overrun with the thundering hoof-beats of yuppies in mass migration' (Milroy 2005:R28).

With the appropriation of two prominent nineteenth-century railway hotels, the Drake and the Gladstone Hotels, and their restoration over the past seven years, the neighbourhood has turned into one of the hippest and trendiest places for clubbers from the suburban areas. In particular, the exclusive and expensive facelift of the Drake Hotel is often held responsible for the accelerated revitalization that brought the centre of Toronto's so-called bohemia further west. When the first Starbucks moved in and, at the same time, oversized billboards with sales pitches for condos went up, the graffiti writers of this town could not resist and spray-painted on the wall of the coffee shop: 'DRAKE YOU HO THIS IS ALL YOUR FAULT'. Less criticism was directed toward the Gladstone Hotel, where the new owners had attempted to smooth the transition of the displaced long-term, low-income tenants by remaining actively involved in their re-accommodation (Graham and Roemer 2007). Some remnants of its past can still be seen in the clientele frequenting the bar, as urban hipsters fill the space in the evenings while the 'older beer-drinking crowd still shows up with military punctuality at 11 am' (Hume 2005:A3).

It was not long before developers discovered the industrial buildings and warehouses just south of Queen Street (Lehrer *et al.* 2006). From spring 2005 on, three different developers bombarded city staff with one re-zoning application after another, asking for condominium towers up to 26 storeys tall and a significant increase in density. The area that experienced this sudden pressure takes the form of a triangle, with Queen Street to the north, the railway corridor to the south and west and Dovercourt Road to the east (see Figure 15.1). On a total of 2.86 hectares, the developers were proposing six new buildings with heights ranging from 8 to 26 storeys, for a total of approximately 1,300 units accommodating up to 2,000 people. These proposals for the WQWT, as it became known, stood in stark contrast to what was in the New Official Plan, which had designated the area for some intensification, mixed-use zoning, expanded existing employment lands and a regeneration area (City of Toronto 2005a). Because the individual proposals were based on independent visions for each site instead of planning the triangle as a whole, the proposals were seen as violating good planning principles.

The city responded with reports to the three individual applications between June 2005 and January 2006 and commented on the re-zoning of the Triangle Area (City of Toronto 2005a, 2005b, 2006). In its reports, the city took the position that the applications were at odds with the planning framework for the area (City of Toronto 2005c, 2005d). The planning department criticized the height and density of the developments, the lack of public parks and community facilities, the taxing demands on existing infrastructure, in addition to the proposals' inadequately addressing the extension of the local street network. Reference was made to New Garrison Common North Secondary Plan, which requires that an area plan must be prepared prior to any significant development. This stipulation is now part of the new official plan for the City of Toronto, but at that time it was still under deliberation at the Ontario Municipal Board (OMB), and the requirement for an area plan became a major focus for residents keen on seeing the Triangle planned as a whole.

By this time, a part of the community had begun to realize that it was under threat of losing exactly the unique social and physical spaces that made the neighbourhood so liveable. Confronted with the reality of having attracted the interest not only of the urban hipsters but also those who turn real estate into a profit-making machinery, the population around West Queen West mobilized. A battle about the future of this neighbourhood was about to break.

Active 18 comes about

Over the summer of 2005, the city held two community consultations that went unnoticed for the most part. It was not until the fall of 2005 that reactions to the developments began to heat up. A number of residents and business owners were frustrated by the way that the City was dealing with developers who presented proposals that were not only out of scale with the rest of the neighbourhood, but did not seem to offer anything in terms of mixed income and mixed-use. A group of residents largely related to the cultural industries came together with the objective of contesting the development proposals and putting pressure on the city for better planning schemes. Referring to themselves as Active 18, the number representing

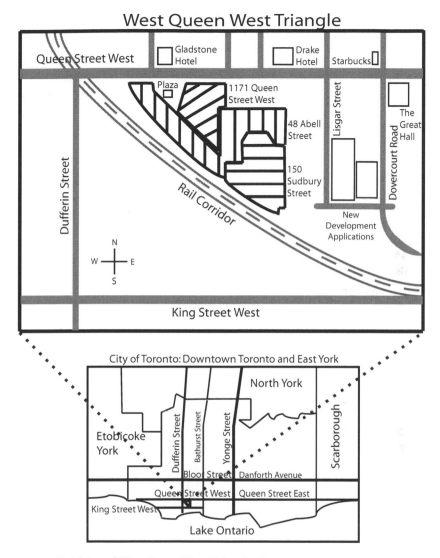

Figure 15.1 Map of West Queen West Triangle, Toronto.

Map by Ute Lehrer.

the Ward in which the WQWT is situated, they held a well-attended public meeting at the Gladstone Hotel in October 2005. Most of the 80 or so participants were disturbed by the massive developments proposed by the three developers who not only wanted to build high-rise condominium towers but had plans to tear down the historic John Abell Factory, built in 1887, which over the past 15 years had become home to a relatively large number of people associated with the so-called creative industries. The studio spaces were affordable, and while live-work space never was formally legalized, it nevertheless was tolerated by the landlord, the same landlord who later proposed to tear down the old factory and replace it with two condo towers. The proposed interventions galvanized the community and led to an elaborate fight against the massive transformation of the Triangle by condo-developers.

This meeting was the formal beginning of Active 18 as an organization representing the interests of the community in negotiations with the city and the developers. It also provided a platform for citizens' engagement in the planning process, even if it was from a relatively remote position. In addition to the public's inclusion in the planning process, Active 18 was also interested in pursuing a collective vision for the development of the Triangle, including the development of a comprehensive Area Plan. Active 18 quickly developed into a well-organized and informed group that was determined to influence development in their neighbourhood. They used their assets effectively; a large number of members were very well connected to planners, architects, media people and other creative thinkers, resulting in much attention not only within the immediate neighbourhood but throughout the entire city.

To this end, Active 18 established itself as a serious community-based group and their efforts in developing their own vision for the Triangle in increasing detail became recognized in the media. Well-respected journalists in the city started paying attention. John Bentley-Mays writes that Active 18 has 'an important point to make' with regard to the historic roots of Queen Street West, and that these roots do indeed warrant 'mindful preservation' and development that considers the character of this 'historic thoroughfare' (Bentley-Mays 2006). By November 2005, Active 18 had set in place mechanisms to develop their own guiding principles for the area. Active 18 networked with other groups that had fought condo developments in the past to learn from their insights and connected with Artscape, a not-for-profit organization that has as its mandate to engage in culture-led regeneration, and secures accommodation for artists throughout the city in an effort to respond to their needs for affordable live-work spaces.

By early 2006, the developers of 48 Abell and 1171 Queen had filed appeals to the OMB, followed shortly thereafter also by appeals of developers at 150 Sudbury, and it became clear that an OMB hearing was unavoidable. In order to be considered equally at the hearing between the City and the developers, Active 18 had to adopt a constitution, elect a steering committee and become incorporated, which happened in late January at a member meeting at the Gladstone Hotel. This was an important legal step in solidifying Active 18's rights to be a party at the OMB hearing and allowing their full participation going forward.

Getting involved in the design process

The members of Active 18 pushed for a collective vision and for better urban design and architecture from the very beginning. The need for a comprehensive approach led Active 18 to get involved in the planning process. In late February, a selected group of experts met with members of Active 18 and discussed the

Triangle with regard to the proposals. The outcome of this pre-charrette was seven planning themes: built form, designing for good retail, affordability, heritage, arts and culture, public space and sustainability (Active 18 2006). These themes were the basis for the Community Design Charrette that was held in March 2006 and lead by Ken Greenberg, a well-respected planner and designer. The first half of the charrette was given to professional experts sharing their insights on gentrification, sustainable development and affordable live-work spaces for artists. In the second half of the charrette, the nearly 100 attendees split into groups discussing the seven themes that had been identified in the pre-charrette. To conclude, the ideas were brought together and it became clear that there were some very strong and tangible outcomes.

After intense discussions within the small groups, eight demands emerged: (1) the need to keep the historic 48 Abell building and its over 100 live-work places intact as the anchor for the new development; (2) to respect the scale and nature of Queen Street; (3) to ensure the redevelopment would be mixed-use zoning and would provide shelter for a variety of income levels; (4) to introduce a substantial amount of public streets and pedestrian walkways to the site; (5) to create multi-purpose green spaces; (6) to link the site with the area to the south through a pedestrian and bicycle bridge; (7) to use a bold sustainability strategy; and (8) to produce a high-quality urban design. The charrette was important in helping solidify Active 18's vision for the Triangle and for establishing the group as an advocate for inclusive planning processes and good urban design.

Being media savvy, Active 18 held a press conference on 30 March 2006, just one hour ahead of the official launch of the Westside Lofts for the proposed condo development on 150 Sudbury (see Figure 15.2). The press conference had three main goals: to draw attention to the developers, who were marketing the condominium units long before they had the approval for redevelopment from the city or the OMB; to call on a moratorium on any new development until the City had approved an Area Plan; and to present Active 18's vision for a healthy redevelopment of the WQWT. Spokespeople for Active 18 used their press conference to make it clear that they were not against development per se, but that they expected an intelligent and comprehensive approach to the redevelopment well beyond the usual profit-making rationale used for most projects. To a question raised during the press conference – how were Active 18 different from the normal NIMBY groups who were not opposed to development per se, but just not in their backyard – Jane Farrell, one of the key spokespersons and the chair of the organization, responded a few weeks later by saying that Active 18 were YIMBYs, meaning that they are pro-development even in their backyard, but that it had to be good development.

The long-winding road of negotiations to (almost) nowhere

In spring of 2006, the so-called working group meetings commenced. Their purpose was to avoid, if possible, an expensive and time-intensive hearing at the OMB. Participants at these meetings were the developers, the City, Active 18,

Figure 15.2 Sale office for Westside Lofts, Toronto.

Photograph by Ute Lehrer.

and some associations from the neighbourhood as well as public agencies that had a direct interest in the WQWT. The general public was allowed to attend but was not authorized to express their opinion. Hosting of the meetings alternated between the City, the developers and the community.

At these meetings a number of issues were addressed, including affordable housing, live-work space, building design, parkland, and historical designation. The city presented its vision of extending the local streets network with connections for pedestrians and cyclists as well as its perspective on the height of buildings. The developers focused mainly on arguing for the necessity of their tall buildings, and to a minor degree addressed the open spaces questions. Active 18 used the meetings to not argue against the development per se but against the form of it. They demanded a comprehensive approach based on good planning principles that would create a mixed-use neighbourhood where green spaces would not solely be spaces between buildings, and where social mix is guaranteed by providing affordable housing and live-work spaces.

At the final working group meeting, Active 18 took the floor to present a summary of the area plan developed in consultation with others. Its main points include development that respects and maintains both the arts and industrial uses of the Triangle; design control for buildings, parks, roads with community input; high environmental standards for building design and construction; the enforcement of a maximum height on Queen Street West; and the encouragement of affordable and family units in new developments.

The working group meetings did not avoid a very expensive and time-consuming OMB hearing in the fall of 2006 where the three developers, the City and Active 18 were arguing for and against each other's perspective. When in January 2007 the decision was made public it was a big surprise to everybody except for the developers: the OMB had supported the proposal of the three applicants in almost all aspects. Margie Zeidler, a founding member of Active 18, summarized the consequences of this decision in a semi-public email: 'Queen Street West will now become a bunch of condo towers ... even if artists could afford to stay they will not – because it will be soon become a crummy mono-culture'. And Charles Campbell, the lawyer who had represented Active 18 on a pro bono basis during the hearings, commiserated that they were 'completely unsuccessful' and wondered out loud 'You have to ask, seriously, why should community groups put in so much effort just to be brushed aside in this fashion' (Campbell 2008).

A final attempt at overturning the decision for a mediocre and compartmentalized development scheme was when the City, with Active 18 as a respondent, appealed the decision of the OMB to the Court. The hearings were scheduled for 16 July 2007, but the evening before the court case was heard, the City, behind Active 18's back, made a deal with two of the developers. The benefits were relatively little.

Conclusion

The story of the WQWT is still unfolding but the important decisions have been made. The developers can go ahead with their massive redevelopment of the Triangle with only minor concessions towards a more comprehensive planning approach. The concerns that Active 18 articulated about the rapid gentrification of the neighbourhood as stated in the quote at the beginning of the chapter, might come true first through the construction sites and their activities that will disrupt the area around the Triangle, and then through the socio-economic transformation of the neighbourhood thanks to the influx of young and hip condo-owners with relatively high disposable income.

There are some small gains. While the developers were not interested in engaging with Active 18 before or during the OMB hearings, after their win, two out of the three made agreements to set aside a certain number of units for live-work spaces for artists. They also took an interest in the design of the open and green spaces and participated in a charrette that was held by Active 18 in March 2008.

After years of being treated as a case-by-case story, the condominium boom in Toronto has reached a state where public debate focused on the future of Toronto's entire built and socio-economic environment. The WQWT is the first case where developers not only faced opposition from the inhabitants and to some degree the City, but had to deal with public debate that reached well beyond the geographic location of the subject site. Having undergone several rounds of contest, it can be assumed that this case will influence planning for the rest of the city for the next few years, and that West Queen West will form a precedent for the new Official Plan and the future of Toronto. The story of the WQWT is unique in many ways but it is also part of a larger narrative of cities competing in the global economy. For the benefit of good planning, let us hope that Toronto's urban renaissance strategies will continuously meet resistance.

Acknowledgements: Funding for this research has been provided by Social Science and Humanities Research Council Canada, Standard Grant, # 410-2005-2202, 'Urban images, public space and the growth of private interests in Toronto'. The research assistance of Kelly Fisher, Andrea Winkler, Laura Hatcher and Constance Exley and the students from the 'Planning in Toronto Studio' May 2006, as well as editorial help by Chris Ouellette and Ian Malczewski are acknowledged.

16 Gentrification and community empowerment in East London

Claire Colomb

Hoxton was invented in 1993. Before that, there was only 'Oxton, a scruffy no man's land of pie and mash and cheap market-stall clothing, a place where taxi drivers of the old school were proud to have been born but were reluctant to take you to. It did not register so much as a blip on the cultural radar. Hoxton, on the other hand, became the first great art installation of the Young British Artists: an urban playground tailor-made to annoy middle England, where everyone had scruffy clothes and daft haircuts and stayed up late, and no one had a proper job. By the end of the 90s, Hoxton had spawned an entire lifestyle: the skinhead had been replaced by the fashionable Hoxton fin as the area's signature haircut, the derelict warehouses turned into million-pound lofts. As the groovy district du jour, Hoxton had come to represent the cliff face of the cutting edge, and everyone wanted a piece.

(Cartner-Morley 2003)

I've heard stories about an up-and-coming area. I mean, for them, not for us. It don't improve it for us at all. There ain't nowhere for kids to play. Just 'cos we got no money doesn't mean we've got no common sense.

(Terence Smith, builder, 41, has lived in
Hoxton all his life, in Taylor 2005)

Locals feel like it is two different worlds down here.

(Goldman, in Taylor 2005)

Responses to gentrification in a 'high-growth', 'housing-scarce' global city

London's processes of gentrification have been documented widely over the past decades, since the invention of the term by Ruth Glass in 1963 in reference to the North London area of Barnsbury. Hamnett (2003) and Butler (2003) have provided deep insights into the dynamics of gentrification in the city, in particular the interplay between economic, demographic, housing-related and cultural processes. While there has been some debate about the extent to which gentrification has led to large-scale displacement, or 'replacement rather than displacement', of the working-class in London (Atkinson 2000a, 2000b; Hamnett 2003), it is clear that over the past 20 years the evolution of the London housing market has made it increasingly difficult for lower- and middle-income groups to retain the possibility of decent living in the city, in particular in its inner boroughs. These income groups include the remnant 'old' working-class, the lower-income groups of the service economy, that is, those employed in a wide variety of service industries 'which do not form part of the classic Blairite image of the new economy', such as clerical, sales and catering staff (Amin *et al.* 2000:22), and increasingly, middle-income workers in the public and private sectors.

When assessing the situation of contemporary London, it seems that any attempt at controlling gentrification must be a lost cause. The housing market is one of the most inflated in the world, due to a combination of shortage, speculative buying, sprawl containment policies, population growth and changing household forms. Following the 'post-recession gentrification' of the 1990s (Lees 2000), London house prices have never been so high: in 2007, the average house price in the city reached £300,000, ten times the average household income. This growth does not appear set to stop and gentrification processes are reaching new frontiers, in particular in the eastern part of the city under the impetus of the 2012 Olympic Games. How, then, can a form of social and functional mix be retained in a 'high-growth/housing-scarce' global city like London? How can community-led strategies be devised to mitigate the negative impacts of rapid gentrification?

Possible responses can come from state intervention and public policies at national or local level (rent control, affordable housing quotas, land-use designations), from community and grass-roots mobilizations (ranging from protest against specific developments to cooperative housing forms) or from a mix of both. One of the early examples of community mobilization aimed at retaining social and functional mix in an area of London undergoing rapid change was the Coin Street development. From the mid 1970s onwards, local community groups mobilized against plans for a large-scale office development south of the River Thames and prepared alternative plans including social housing, a park, managed workshops and leisure activities. In 1984, the 'Coin Street Community Builders', a development trust and social enterprise set up by local residents, managed to buy the site and gradually implemented their mixed-use vision for the area. Affordable housing for low-income workers has been cross-subsidized by the profits generated from commercial leases on the riverfront. The Coin Street 'success story' has been widely documented (Tuckett 1988; Baeten 2000; Brindley 2000), but it remains the exception rather than the norm in London's recent neighbourhood regeneration history.

This contribution explores the (small) room for manoeuvre in contemporary London for local actors to mobilize and regain control over the impacts of gentrification on their neighbourhood and work towards more equitable urban redevelopment outcomes. It focuses on one example of mobilization in Hoxton/Shoreditch, East London – an area which has undergone rapid gentrification in the 1990s. The area is adjacent to the northeastern edge of the 'Square Mile', the heart of the financial and banking industry in the City of London. It has consequently been under intense pressures for (re)development since the 1980s to accommodate new service activities, new businesses and high-income residents. Responses to the rapid gentrification of the Hoxton/Shoreditch area are explored through a short case study of a Local Development Trust (LDT) which has worked towards reclaiming 'community spaces' amidst soaring property values.

The context: the gentrification of 'ShoHo' in the 1990s

The areas of Shoreditch and Hoxton (nicknamed 'ShoHo'), part of the Borough of Hackney, form the northern part of the historical East End of London, the cradle of successive waves of immigration throughout the city's history – Huguenots, Irish, Jews, Bengalis. The 'city fringe' boroughs of Tower Hamlets, Hackney and Islington ranked fourth, fifth and sixth most deprived areas in the United Kingdom in terms of their average deprivation score (Shaw and McLeod 2000). The Shoreditch and Hoxton areas thus suffer from high unemployment and low economic participation rates despite the fact that the number of jobs on the city fringe is four per economically active resident. Nearly 40 per cent of the local population comes from an ethnic minority background.

The East End has historically been home to activities not welcome within the city's wall (industries, theatres, fairs) (Shaw and MacLeod 2000). From the Victorian era to the 1980s, Hoxton and Shoreditch were working-class neighbourhoods characterized by small industries and workshops in the furniture and shoe trades, thriving street markets, music halls and cinemas. The area was heavily bombed during the Second World War and a lot of council housing was subsequently built to deal with housing shortage. Hoxton became notorious for organized gang crime. Unlike other parts of London which were gentrified early on, Hoxton's built environment is not dominated by Georgian and Victorian terraced houses, but by workshops and former industrial buildings. From the 1980s onwards, it is precisely this type of building stock which attracted artists looking for cheap 'live and work' space. The old building stock is interspersed with 1960s to 1970s social housing blocks, owned and managed by Hackney Council or by housing associations.

In the early 1990s, the area began to undergo a significant change as artists and curators on the lookout for cheap studio spaces began to move around Hoxton Square. The commercial art gallery White Cube was opened in 1999 in a former factory. The area became the epicentre of the Young British Artists scene, among them Tracey Emin and Damian Hirst. Several galleries opened around Rivington Street, Curtain Road, Charlotte Road and Hoxton Square. The concentration of artists and art happenings then drew other creative professionals into the area – film makers, architects, designers, musicians, actors and media professionals. This kick-started a typical gentrification process. Bars, cafes and restaurants soon followed, turning the area into a popular night life district packed with large crowds of revellers on weekend nights. By the mid 1990s, United Kingdom and international newspapers were labelling the area as the 'hot spot to be' (Cartner-Morley 2003). After the young creatives, came the developers, eager to capitalize on the image of the area and its proximity to the City of London. Eventually came the 'celebrities' ...

As a result, the area underwent a significant rise in real estate values and many of the original artists were priced out less than ten years following their settlement. Other artists have deliberately left the area as they felt that 'popularity has killed personality' and that the night economy has altered the 'exclusive' and 'edgy' character of the initial artistic community, as argued by fashion editor Cartner-Morley (2003). The planned extension of the East London Underground Line (due in 2010) provides an additional impetus for the continuing gentrification of the area, now spilling eastwards towards the Dalston and Bethnal Green areas. The story of Hoxton/Shoreditch thus resembles very closely the processes analysed by Zukin in her account of the transformation of SoHo in New York (1982):

> enclaves of art and craft production emerge in areas of low-rent accommodation, notably redundant industrial buildings and warehouses, which provide studio and living space for people with creative talent but little money. Later, such places acquire prestige with young professionals – a fashionable address which confers status through association with creativity. Rising land values displace both working class residents and artists. Furthermore, established communities and businesses may feel that they are losing 'ownership' of the public places and facilities in their immediate neighbourhood.
>
> (Cited in Shaw and MacLeod 2000:166)

The contrast between the 'old' and the 'new' Hoxton is striking, as illustrated in Figure 16.1. Physically, in spite of the geographical proximity between different housing types – often in the same street – there is a quasi-complete separation between the refurbished loft apartments and newly built secure developments on Hoxton Square and Kingsland Road and the social housing estates located a few blocks away. Few visitors venture beyond the Square and realize the extent of deprivation and grime in the 25 housing estates in the area. While a number of estates have been improved through various government funding schemes, some are still in a very poor state.

This micro-geography of housing segregation is reinforced by the separation between the spaces and practices of socialization of old and new residents. As in other gentrified neighbourhoods in London, there is 'something of a gulf between a widely circulated rhetorical preference for multicultural experience and people's actual social networks and connections' (Robson and Butler 2001:77). Socially, the contrast between the 'old' residents (mostly social housing tenants), the 'Hoxton trendies' (pejoratively nicknamed 'Shoreditch Twat' – local slang for 'a new media, fashion student, photographer-type person with a privileged digital or old school arts background who lives/works/socialises in London's East End area of Shoreditch', Urban Dictionary 2007) and the most recently arrived high-income professionals is striking. The Hoxton area epitomizes the mosaic of 'utopian and dystopian spaces', 'physically proximate but institutionally estranged' (MacLeod and Ward 2002:154) which is characteristic of many of the recently regenerated areas in UK inner cities.

Figure 16.1 Contrasted urban landscapes in Hoxton/Shoreditch, London. Clockwise from upper left: social housing block, derelict cinema, refurbished loft apartments, graphic design firm.

Photograph by Claire Colomb.

The price paid by the existing local communities as a result of the rapid changes in the neighbourhood is high. Most locals do not benefit from new jobs being created or are confined to low-paid bar or security work (City Fringe Partnership 2003). Children find it difficult to stay in the area when they move out of their parents' home. Even if the high density of social housing has mitigated gentrification-led mass displacement (thanks to stable tenancy agreements), there have been clear conflicts over the use of green, cultural, social and retail spaces and a widespread feeling of alienation among 'old' residents (Goldman and The Red Room 2005; Taylor 2005). New bars and restaurants are too expensive for 'old' residents to use. The large crowds of revellers taking over Hoxton Square at night create major disturbances and unsafe environments for locals. New cultural facilities, such as the Circus Space, are scarcely used by local residents. In Hoxton, The Crib, a drop-in youth centre used by vulnerable children, was closed by Hackney Council in December 2004 to make way for flats. The centre's closure generated anger in the local community (Taylor 2005).

Local residents' perceptions and feelings about the changes in their neighbourhood are therefore very mixed (Taylor 2005), reflecting the ambiguous nature of inner city revitalization: one person's regeneration is another's gentrification. This was well illustrated by a project set up in 2005 by local artists based in the Hoxton area, called the 'Hoxton Story'. The project aimed at exploring Hoxton's 'mythical' regeneration from the point of view of the people who live there, through a visual and oral montage of facts, fictions and verbatim testimony. It juxtaposed the perspectives of tenants, squatters, residents, council workers, community activists, artists, architects and developers, and 'explored their dreams, disappointments and achievements' in relation to the environment of Hoxton 'as home, work and recreation; as public space and as capital waiting to be realised' (Goldman and The Red Room 2005). The project was realized by The Red Room, a radical theatre project founded in 1995 to address political, social and cultural issues through performing arts. The project illustrates the local engagement of a small number of 'first generation' Hoxton artists who settled in the neighbourhood.

The Red Room's director, Lisa Goldman, interviewed more than 30 local people to write the script for the 'Hoxton Story' and create an oral history audio-archive of residents' voices about the changes in their neighbourhood (Goldman and The Red Room 2005). The project ended with a series of participatory theatre plays, or 'intimate walkabout performances', in which local residents and professional actors jointly took the audience in small groups through the streets of Hoxton – from a surviving music hall through the local market to council flats and shabby tower blocks, green spaces, a community centre and a trendy bar on Hoxton Square. At the end of the performance, the audience was carried in a hearse to attend the (metaphorical) funeral of Hoxton. The performance blended personal testimonies with fictional narratives and characters to draw the audience into an intimate, emotional vision of the contrasting perceptions of change in the area, asking who benefits (or not) from those changes.

The Shoreditch Trust – fighting for community spaces 'on the back of gentrification'?

Within the context of the rapid gentrification of the area, a local partnership scheme called 'Shoreditch Our Way' (ShOW) was set up in 1999 to empower local residents. The ShOW initiative was set up within the framework of the New Deal for Communities, an urban policy programme launched by New Labour in 1998 and implemented in 39 deprived UK neighbourhoods. This area-based programme is characterized by a new political language of social inclusion and 'people-based regeneration'. The ShOW partnership received £50 million over ten years to implement regeneration measures covering an area of 20,000 residents, 60 per cent of which were council or housing association tenants. Although the ShOW partnership was set up initially as a vehicle for the implementation of a central government-funded urban policy initiative, it drew from an established collaborative network of tenants' and residents' organizations which had been in existence in the area prior to 1999 (Perrons and Skyers 2003:277). The partnership later worked towards gaining long-term autonomy from government funding. In 2004, ShOW was renamed the Shoreditch Trust.

The Shoreditch Trust is a not-for-profit LDT, a type of organization which has gained increasing popularity in the United Kingdom as a vehicle for community-oriented urban regeneration. LDTs are non-profit entities whose membership is drawn from a geographically defined area and whose board is made up of representatives from the public, voluntary/community and private sectors who may be elected from 'voting sections' of the local community (Development Trusts Association 2007). Funding may come from the public sector, from charitable grants and from the profits generated by subsidiary social enterprises. Many LDTs have turned towards urban development activities by buying and managing land and buildings as assets for their organization and for the local communities they are serving.

The sophisticated structure of the Shoreditch Trust ensures a good representation of local residents into the decision-making process establishing economic development and regeneration priorities (Perrons and Skyers 2003:277): the Trust is managed by a board of 12 members elected by local residents and 11 representatives from the public, private and voluntary sectors. Elections take place every three years and innovative techniques are used to generate interest in the vote among local residents, for example, by organizing voting sessions in schools. Since 1999, the composition of the board has changed significantly, from a majority of white, working-class women, to a mixture of local youth, ethnic minority residents, social housing tenants and 'new' upper-middle class professionals recently arrived in the area (Pyner 2007).

The work of the Shoreditch Trust, which now employs 28 people, has been multi-faceted: training and employment measures, capacity-building activities, cultural and youth projects, physical improvements to housing and infrastructure including a community transport project (Shoreditch Trust 2007). One of the most interesting aspects of the work of the Trust is its deliberate use of some of the funding received from central government to 'buy up buildings before commercial developers move in' (Taylor 2005), thereby putting together and managing a stock of properties which can be retained for various community purposes. The Trust has thus directly addressed some of the concrete conflicts over the use of space brought about by the rapid gentrification of the area.

One of the Trust's flagship projects is a former Victorian school located at 16 Hoxton Square, converted in 2004 into a community centre providing learning, personal development and enterprise opportunities for local residents (see Figure 16.2). The Trust owns the building but devolved its management to a non-profit training charity. A social enterprise restaurant, the Hoxton Apprentice, provides catering and hospitality training for local homeless and long-term unemployed. All profits from the restaurant are reinvested into the charity. The building also houses business incubator units, a community sports centre and meeting facilities. The top floor of the building was converted into two luxury penthouse apartments sold by the Shoreditch Trust under the English leasehold system. The profit generated by this sale was used to cross-subsidize some of the community-oriented

Figure 16.2 16 Hoxton Square community centre, London. Retaining spaces for local communities amidst upmarket bars and art galleries.

Photograph by Claire Colomb.

activities in the building. The Trust's chief executive pointed out how 'what could easily have become another luxury development beyond the reach of ordinary residents is now a community resource' built for, and by, the people of Shoreditch (16 Hoxton Square 2007).

In another Shoreditch location, on the site of a disused clothing factory by the Regent's Canal, the Trust, through its property arm (the Shoreditch Property Company), has bought part of a privately developed building, the 'Canalside Works'. The Trust rents out 2,000 square metres of workspace at affordable rates and flexible lease terms for small enterprises who currently struggle to find appropriate space in the city fringe. The building houses a social enterprise eco-cafe owned by the Trust. The Trust has also worked with the Metropolitan Police and Hackney Council to co-locate key neighbourhood enforcement and community safety services in the building – a pioneering initiative. The remainder of the development hosts private housing and commercial uses. The mix of functions and residents seems to work well and without tensions (Pyner 2007).

The Shoreditch Trust has thus acted as a developer, acquiring buildings and mixing commercial and community uses in its properties. Additionally, private developers have in some cases approached the trust to 'offer' working space as part of new developments being built in the area. The English planning system allows conditions or agreements to be negotiated between local authorities and developers to secure the provision of affordable housing, workspace or community facilities. Such obligations are attached to the planning permission granted to a private development scheme. Some developers have fulfilled their obligations by giving out workspace in newly developed buildings to the Trust at no cost. As of 2007, the Trust owned £7 million worth of real estate assets. The building stock accumulated by the Trust, however, does not generate a very high income. The Trust's activities have thus been complemented by the development of a number of 'social enterprises', that is, companies with charitable status. One of them, ACORN House, a successful eco-restaurant located in the King's Cross area, has paved the way for similar venues across London. By 2011, the Trust foresees that £1 million per year will be generated through these social enterprises, which can help support the day-to-day running of the Trust and secure its longevity after government funding stops in 2010.

The role of the local state in the story of neighbourhood change in Shoreditch/Hoxton has been ambiguous. On the one hand, Hackney Borough Council has supported the formation of the ShOW partnership and the activities of the Shoreditch Trust (whose activities remain, however, independent from the Council). On the other, it played an indirect role in some of the gentrification processes witnessed in the neighbourhood, for example, by selling public buildings for commercial uses or by encouraging the settlement of creative industries at all costs in specific sectors such as fashion, product design and the visitor economy. Although efforts have been made to target local unemployed residents, most of these economic sectors are the traditional engines of neighbourhood gentrification – one of the contemporary paradoxes of 'marketing the creative industries' in deprived neighbourhoods as an urban regeneration strategy.

Conclusion

Drawing conclusions on the extent to which the Shoreditch Trust represents a successful form of community control over the area's urban redevelopment is not easy. Some of the constraints in that respect are the degree of autonomy of the Trust within the parameters set by central government-funded programmes, and the limited degree to which community participation can influence material resources allocation, within and outside the area (Perrons and Skyers 2003). It can nonetheless be argued, with regard to the particular problematics of gentrification, that the Shoreditch Trust's activities have represented a positive form of local mobilization against the negative impacts of rapid gentrification on social and community spaces. The director of the Trust describes the strategy as 'using gentrification through the front door'. For existing residents and businesses confronted with processes of neighbourhood change triggered by external forces, he argues, the choice is twofold: a passive, adversarial and negative attitude – rejecting neighbourhood change en masse and facing the risk of being mere 'victims of change' or a more proactive, positive attitude aiming at harnessing opportunities arising from the transformations of the neighbourhood for the benefit of local communities. The Shoreditch Trust has thus tried to move 'community activity' beyond a confrontational style:

> Newcomers can be perceived as the enemy, but they are also those who bring new opportunities. The 'politics of envy' between low income residents and new higher income residents can be shortcircuited through visible improvements in education, healthcare, housing, leisure and cultural facilities for existing residents.
>
> (Pyner 2007)

The lessons from this 'Hoxton Story' do not come as a surprise when compared with the analyses of previous initiatives such as the 'Coin Street Community Builders'. What made a difference to the fate of the areas concerned, in terms of the retention of social and functional mix in a context of rapid gentrification, is the fact that organized community groups gained ownership and control of land and/or buildings, and were able to implement their vision of a mixed-use neighbourhood (albeit at different scales in the two areas). The starting point was the firm belief that existing residents should be maintained in the area and that change should be harnessed in a positive way to improve their living conditions. Both organizations found mechanisms to cross-subsidize community-oriented activities on the back of increasing land values in the area. In that context, the primary rationale for the Shoreditch Trust to act as 'developer' and 'entrepreneur' through property development and social enterprise activities has been twofold: bringing direct benefits to existing communities by allowing the retention of key spaces for community uses, and securing some longevity for the organization and a higher degree of independence from government funding by building a stock of real estate assets (even if the income generated may remain limited).

In both cases, however, support from the state was also instrumental to the process: in Coin Street, the local state ceded some of the land to the community group; in Hoxton/Shoreditch, a large sum of central government funding helped support the foundation of the Trust and the initial purchase of buildings. This points towards the paradoxical and contradictory role of the state in contemporary processes of gentrification and neighbourhood change (Jones and Ward 2004; Colomb 2007) – many of its policies are part and parcel of a neoliberal urban project which facilitates or even encourages gentrification as an instrument of urban renaissance (Atkinson 2004), but at the same time some of its agencies provide funding, instruments and legislation which can be used by local groups – among other instruments – to address the adverse impacts of neoliberal political and economic restructuring on the inner city, albeit to a limited degree. Initiatives such as the Shoreditch Trust alone cannot reverse the trend towards wholesale gentrification in a city like London, in which the housing problem is rooted in regional, national (even global) trends and policies. They can, however, help retain key social infrastructure for lower-income groups in affected neighbourhoods, and pioneer new forms of community empowerment in neighbourhood regeneration.

Part IV

On the possibilities of policy

Gertrude Street Fitzroy

Picture by Rodger Cummins,
text by Kate Shaw

Gertrude Street Fitzroy.

Gertrude Street Fitzroy is in the heart of one of the first districts in Melbourne to gentrify. It borders a large public housing estate, and is home to a relatively stable population of early gentrifiers and a long-standing Aboriginal community. The street contains drug and alcohol services, junk and antique shops, reduced-price bookshops, and some of the funkiest bars in the city. This is where Bob Dylan comes when he's in Melbourne and not being seen. How long can a place remain in this ambiguous, transitional state? For Gertrude Street, so far, about 40 years.

The sign reads:

"The Victorian Aboriginal Health Service (VAHS), a Fitzroy based Aboriginal community controlled organisation established in 1973, has a long-term lease for this building. The VAHS operated out of this building between 1979 and 1992 after moving from its first site at 229 Gertrude Street. In 1992 the VAHS moved into a purpose built facility in Nicholson Street [around the corner].

In late 2006, a public tender process was undertaken by the VAHS to identify a business to sub-let the building.

Mission Australia was selected and proposes to run a social enterprise cafe and related training facility from the building as part of its youth transitions program. It will be an inclusive and catered meeting place for people to experience Aboriginal culture and foods.

A key outcome of the project will be to provide employment, training and education opportunities for Aboriginal people in the retail and hospitality industry.

The rental income will be used by the VAHS to resource health programs for the Aboriginal community. Any profits made by the cafe will be used by Mission Australia to support the youth transitions program.

Major works are being undertaken to the building with the support of the Victorian Government. These works are expected to be completed by late 2008."

17 Heritage tourism and displacement in Salvador da Bahia

Elena Tarsi

Translation by Bonnie A. Rubins

> *Toda a riqueza do baiano, em graça e civilização, toda a pobreza infinita, drama e magia nascem e estão presentes nessa antiga parte da cidade.* (All the riches of Baiano, grace and civilization, all the endless poverty, drama and magic are born and are present in this old part of town.)
>
> (Amado 1977:67)

The city of Salvador in the state of Bahia is an important landmark in Brazil. With its 450 years of history, it is one of the oldest urban colonizations in Latin America, the first capital city of Brazil and the second of the Portuguese Empire. Its historic and cultural richness gives Salvador a strong identity in the process of contemporary transformation. In this chapter, I will reconstruct the long functional transformation that for four and a half centuries has taken the city's historical nucleus, the Pelourinho, from its primitive splendour to social and aesthetic deterioration. I will analyse the public policies that, from the 1970s onwards, have made it the object of physical regeneration and symbolic investment.

I will show how regeneration policies were based on exploitation of the historic centre's tourism potential without considering the importance of the social fabric. The safeguarding of heritage only involved the conservation of old buildings and churches without considering the immaterial heritage, the popular culture, which, instead of being safeguarded, was expelled along with the inhabitants. This case is emblematic of both urban interventions executed in Salvador and to which values priority had been given, with no consideration for the rights of the population directly involved.

However, in recent years, there has been a substantial change in policy objectives concerning the historic centre. Thanks to the popular struggle in defence of poor people's 'right to the city', carried out primarily by Afro-descendents, the policy of safeguarding and regenerating the historic centre has been integrated with a social policy allowing old residents to remain in the Pelourinho, affirming the right of the poor to the city.

Historical background

In 1985, the historic centre of Salvador joined the UNESCO list of World Heritage Sites as one of the greatest Baroque legacies outside Europe. The original nucleus of Salvador was born in the area overlooking the current port, along the cliff that separates the site into two sections. Activities associated with the maritime trade took place in the Cidade Baixa on the narrow coastal plain, and the Cidade Alta sat on the small plateau about 80 metres above sea level. The city's upper part was the administrative seat of the colony until 1763, the year in which the Imperial capital was transferred to Rio de Janeiro. The first settlement began to grow quickly in response to the economic and commercial development of the colony, an 'original Brazilian place composed of Africa, created in Europe and located in America' (Memoli 2005:15). The city, which was shaped from the late sixteenth century onwards by the sugar economy, was the centre of the Recôncavo, the region that surrounds the Baia de Todos os Santos, and had the largest urban network created in the Americas by a European colonial power. It became a cultural centre, the religious art and architecture of which reached very high levels. The enormous wealth that was concentred in Salvador, especially from 1650 to 1800, came from the indiscriminate exploitation of local resources and the labour of slaves deported from Africa.

All of the city's activities concerning economic and administrative control as well as its cultural and political life were concentrated in the Pelourinho, where the Recôncavo landowners resided until the beginning of the past century. The Pelourinho was the absolute centre of the city until mid-century, when functions began to be redistributed to the nearby areas.

The word Pelourinho refers to a stone column located in the centre of a town square, where criminals were put on display and punished. In colonial Brazil, it was primarily used for the whipping of slaves. Amado described the square where the column once stood as follows: 'the paving stones are black like the slaves that sat on them, but when the midday sun shines more intensely, they reflect the colour of blood' (1977:67). The largest negro (the term, as it is used in Brazil, indicates the concept of belonging to a common culture of African origin) city outside of Africa, Salvador inherited the resistance fought by slaves, who had organized a complex network of institutions and communities, creating a matrix that left a deep imprint on the Brazilian character and identity. Resistance and historic memory have produced a strong, underlying affirmation of the Afro-descendent culture: from the formation of communities of escaped slaves, called *quilombos*, through the struggle for independence and the urban insurrections of the nineteenth century, to the ongoing battle for racial democracy.

Pelourinho: the deterioration

In the early twentieth century, the city began a linear expansion that first lengthened the north–south corridors of the historical nucleus and later filled in the remaining empty spaces. The upper-middle classes began leaving the centre districts and moving to nearby neighbourhoods (Vitoria, Gracia and Barra) that were better adapted to their new standards of mobility. This led to a radical transformation of the Pelourinho. The void left by the upper classes was filled by popular commerce and various types of services, to which was added a new temporary resident population made up of rural immigrants, students and foreigners. The greatest Bahian writer, Amado (1999), in his novel *Sudore*, tells stories of people living in a building in the Pelourinho, giving us a colourful view of the social climate in the historic centre in the mid 1900s.

In only a few years, the historic centre of Salvador began to suffer a rapid loss of traditional functions, decay of buildings and deterioration of urban quality, accompanied by extreme poverty, a large degree of social mobility and marginalization of residents. 'In the 1930s, the municipal government decided to concentrate prostitution in the Pelourinho quarter' (Memoli 2005:65), freeing up the space where it had been historically located outside the walled city and allocating this area to commercial functions.

The road infrastructure built from the late 1960s onwards completely altered the volume of vehicles and persons in circulation, to the extent that the historic centre became marginal with respect to the more dynamic processes of transformation of the urban structure. Various public organs were progressively moved from the centre, while a deliberate policy that favoured investments away from the consolidated area led to the stretching of the urban fabric and the creation of new business centres, high-quality residential areas and large quarters in distant areas. Moreover, the moving of businesses to more populated areas, the fragmentation of public activities and the reduction of the economic role of the port sent the central area into a crisis accentuated by the progressive decay of its physical structures.

This was a common form of evolution in many cities where the abandonment of central areas by the middle classes and the expulsion of the most important functions left the areas to the marginal classes and economic and social roles of little importance. Many South American cities have dealt with the deterioration of their central areas, even large ones, which have become unsuitable for circulation, by demolition and reconstruction.

Conservation policies

This loss of the dynamism of Salvador's old centre coincided on a national level with the redefinition of the Brazilian policy of conservation of its historical and cultural assets. 'On an ideological level, the 1970s sought a broad notion of historical assets in order to build a national identity stimulated by the military regime' (Fernandes and Filgueiras Gomes 1995:51). In 1967, the state government created the Fundação do Patrimonio Artistico e Cultural (Foundation for Artistic and Cultural Heritage) – now the Instituto do Patrimonio Artistico e Cultural da Bahia (Institute for Artistic and Cultural Heritage of Bahia) – whose first initiative was to transform the Pelourinho into a tourist centre. The tourism potential was recognized as vitally important for the economic future of Bahia, a place with a special vocation for a growing market at a global level. In the early 1990s, tourism made up 3 per cent of Bahia's GDP with the possibility of an increment in existing investments. In the municipality's official documents, the Pelourinho was actually defined as the 'most characteristic representation of the city. It is with this that we can sell the city on a national and international level; it is its heart' (CONDER 1996:5).

The initial regeneration project in 1970, the Plano de Cidades Historicas (Plan for Historical Cities), which was aimed at recovering the old cities of the Northeast with tourist potential, was founded on the social revalorization of the district through the deportation of old inhabitants out of the *bairro*. This was to have been followed by public restoration of the only central square in the Pelourinho, in hopes of creating a cascade effect to which the market would react, expanding the redemption to the entire area.

The project was not very successful primarily due to the lack of articulation among the three levels of power. However, it signalled the beginning of a long series of projects that, in the 1970s and 1980s, were presented and launched with very uncertain results.

Many of these were thwarted by both legislative limits that restricted funding only to public buildings and by the lack of an integrated policy on the social marginalization of the area. As a consequence, even private investors did not respond adequately, seeing that focusing only on architectural restoration would not guarantee a renaissance of the area, still dangerous and degraded.

In the meantime, the inhabitants organized a movement in the late 1970s to defend the residents' right to remain in the historic area for the entire following decade. Until the late 1980s, very little else went on except for the regeneration of the most important buildings and the main churches. In other words, the intervention policy produced very fragmented results from an urban and social point of view.

In the early 1990s, a far-reaching cultural movement of great participation opposed the decay of the Pelourinho, taking on various expressions among the people and producing intense cultural activity. This was a true reclaiming by the Afro-descendent culture of the identity of the old centre through the affirmation of its symbolic values, creating a cultural and political movement expressed essentially through ethnic identity. The urban restoration, which focused on traditional identity symbols, developed a slow and continuous process of affirmation of black identity. According to Memoli, 'The large black population became the promoter of an important political and cultural movement that took shape in the late 1980s and was rooted in the unrest that had rippled throughout the Afro-American world beginning in the 1960s and 1970s' (2005:83). The black society claimed visibility, which was manifested in the Pelourinho by an increase in the number of Afro *Blocos* in the carnival. These were almost exclusively made up of black people and became the deepest expression of the desire to renew African traditions publicly giving a voice to the deep, underlying texture of local culture. The associations that had sprung up all over the city, especially in the more popular circles, manifested their existence and held their ceremonies in the historic centre, which was seen as an identifying, significant and traditional space. 'It is the exceptional importance of the centres as communities and social realities which turns them into objects of great symbolic value. Controlling the centre and access to it represents not only a concrete material advantage but also the dominion of all its symbology' (Villaça 2001:241).

A characteristic of the black cultural movement was that it broke with the 'purism' usually linked to such a movement in the sense that, on one hand, cultural production underlined the importance of tradition, while on the other hand, contemporary reinterpretation of tradition fed back into the cultural market.

(Fernandes and Filgueiras Gomes 1995:54)

To this process was added the explosion of tourism and the cultural industries, which was occurring on both national and international levels in those years. Therefore, tourism was expanded as a vocation of the city, along with a powerful movement to affirm black identity and a new articulation of the question of cultural defence and participation in commercial channels for culture.

Harvey notes that while the image serves to establish an identity on the market, it is also the founder of the identity of the city (1997). Memoli extends this:

Being unable to recreate the reality in all of its complexity, the image takes into consideration only parts of the territory, adapting itself to the potential and the language of the means used. The global need to create a strong identifying image also becomes a political means of involving the marginalised strata of the population in a sense of belonging to that image.

(Memoli 2005:22)

This happened in Salvador where the city sold itself on the global level using the strongest expressions of its own traditional culture, which were still part of a battle to affirm an entire culture that was as rich as it was marginalized.

This was the scenario in 1992, when the state government began a restoration operation of the historic district with funding from the Caixa Economica Federal (Federal Economic Bank) and the Banco Interamericano do Desenvolvimento (BID – Inter-American Development Bank). The programme was initially divided into four phases, each of them corresponding to a parcel, to which another three were added later. The two main characteristics of the programme were restoration of entire blocks rather than individual buildings, as had been done previously, and prior definition of the use of buildings after restoration. Once again, a policy of social 'cleansing' was presented: the inhabitants were offered the choice of moving from their homes to suburban neighbourhoods, leaving outright in exchange for compensation, or waiting for a possible but not certain assignment of a renovated home. Considering the economic and social reality of the old residents, 'victims of psychological pressure and without knowledge of their inalienable rights' (Cardoso and Saule 2005:93), many opted for compensation, allowing for the liberation of the quarter and the concession of the spaces to business people interested in establishing themselves in the area, especially owners of tourism services, such as restaurateurs, hoteliers and retailers. Investments were also made in the basic infrastructure (water, electricity, sewers) and the face of the neighbourhood was painted with lively colours to highlight its architectural similarities and create a stronger visual impact compared with that of the recent past and the nearby areas, which were still in a state of deterioration.

With the creation of the Segretaria da Cultura e do Turismo (Secretariat of Culture and Tourism) of the state government in 1995, the programme for regenerating the historic centre fell under its jurisdiction. In addition to concluding the work in progress, it started up a cultural animation project entitled '*Pelourinho Dia e Noite* with the objective of attracting a flow of citizens and tourists through free shows and events' (Barreto and dos Santos 2002:126).

In just two years, by the end of the fourth phase, there had been interventions in 334 buildings on 16 blocks, more than 1,800 families had been evicted and more than 150 small businesses had closed (data taken from Barreto and dos Santos 2002:57). The fifth and sixth phases will proceed similarly. With respect to past projects, this intervention was considered effective for various reasons, including the careful political and urban marketing campaign conducted by the state government and the volume of investments obtained which gave credibility to the intervention. However, the fact remains that the expulsion policy aimed at the old residents, based on lack of recognition of the rights of poor people, had profoundly jeopardized the local development. The social and morphological character of the centre had been injured in the heart of its original vitality and replaced with a 'postcard' to be sold on the global tourism market. Moreover, in this case, the violation of the 'right to a home' and the 'right to the city' is clearly an act of discrimination: poor black people were denied their right to live in the Pelourinho:

> The displacement of inhabitants showed a complete lack of respect for people's rights, as all the social, economic and cultural connections of families depend on the neighbourhoods where they live and work. Moreover, the 'right to a home' for low income people living in the historic centre was totally ignored and there was also no consideration of the fact that this was the place where the poor had representation in the affirmation of the *negro* culture.
>
> (Cardoso and Saule 2005:93)

Monumenta

In 1997, an agreement between the Ministero da Cultura (Ministry of Culture) and the BID led to the creation of a programme entitled Monumenta, which launched a series of projects in the largest Brazilian cities. The programme encouraged the regeneration and conservation of historic assets in urban centres recognized by the Instituto do Patrimônio Artístico e Histórico Nacional (National Institute of Artistic and Historic Heritage), including museums, churches, monuments, roads, palaces, public buildings and even private buildings. Collaboration between Monumenta and the state government, which was responsible for carrying out the programme, began in 2000 with the seventh phase of the restoration of the historic centre: an area of about ten blocks between Praça da Sé and the Elevador Lacerda.

The programme continued to promote the eviction of low-income families, offering conditions to encourage them to either move to the city suburbs or receive financial compensation. The precarious conditions of the homes ensured that the majority of the inhabitants accepted the proposals, in fact, out of 1,674 families; only 103 decided to remain in the area. The families that remained – who earned their living by small tourism businesses, such as *Acarajé*, jewellery, crafts and *Capoeira* – began a movement against the eviction policy which culminated in the creation of the Associação dos Moradores e Amigos do Centro Histórico (AMACH – Association of Inhabitants and Friends of the Historic Centre) and the filing of a civil action with the public prosecutor's office against the state government and the Companhia de Desenvolvimento Urbano da Bahia (CONDER – Company of Urban Development of Bahia).

According to Cantarino:

> These interventions placed an emphasis on the term 'culture', which was a vision of culture with a political dimension that was not concerned with either the people's 'right to the city' or the struggle against exclusion and inequality. Culture is developed on the field of contrast, where even the historic heritage becomes a realm of material and symbolic struggle amongst social groups and values in conflict.
>
> (2007)

The families won the struggle and obtained a radical change of the nature of the interventions in the area involved in the seventh phase of the regeneration project: the main objective finally shifted from the expansion of tourism to the question of housing.

On 29 July 2004, at a meeting that took place in Salvador and which included representatives of both the residents and the relevant authorities (AMACH; Assessoria de Desenvolvimento Urbano do Estado [Consultancy of Urban Development of the State]; Unidade Executora do Programma Monumenta [SEDUR – Executing Unit of Monumenta Programme]; CONDER and the Monumenta Programme), the programme was reformulated. The residents were to be included, represented by AMACH, in the discussions and negotiations. A range of proposals were presented: delineation of the seventh phase of the regeneration project should occur with the community in a participatory format, so that the architectural projects are compatible with family compositions and to consider the residents' desire to remain in the homes where they have always lived; identification of the families living in the seventh-phase intervention area by SEDUR, CONDER and AMACH to guarantee that they would remain in the area after the restoration; moving of families living in the seventh-phase intervention area to temporary housing during the interventions, with the guarantee of both returning to live in the area after the restoration and that their temporary housing would be adequate to allow them continuity in their working lives; involvement of the Segretaria Estadual de Combate a Pobreza (SECOMP – State Secretariat for the Struggle Against Poverty) in the development of seventh-phase interventions that would create work and income opportunities for inhabitants of the historic centre; creation of a permanent office for AMACH in the seventh-phase area; creation of a specialized drug rehabilitation and assistance centre within the boundaries of the seventh-phase area; and revision of compensation amounts offered to residents who opted not to remain in the area.

The result of this negotiation process was the signing of the Termos de Ajustamento de Conduta (TAC – Terms for the Regulation of Conduct), 'which established a series of state government responsibilities in seventeen points to answer to the needs expressed in the Civil Action' (Ministério das Cidades 2007). Following on both the initiatives of federal representatives and social movements and the action plan of the restoration programme for central urban areas, the TAC set up a steering committee with representation from the population and a permanent AMACH headquarters in the Pelourinho.

The agreement included the permanent residency of the 103 families and the setting up of services requested by the inhabitants, such as a community laundry, play areas for children and spaces for community social life. The restoration was not, in other words, to be a mere display. The majority of the inhabitants worked in the historic centre in small businesses or as street vendors. They were given the opportunity to work on the restoration of the buildings through a manual labour training programme.

To carry out the principles expressed by the Public Prosecutor, which corresponded to the action plan in the restoration programme for central urban areas, the Ministero da Cultura (Ministry of Culture), represented by the Monumenta programme, the Ministério das Cidades, through the Segretaria Nacional de Programa Urbano (National Secretariat of Urban Programmes) and the Segretaria Nacional de Habitação (National Secretariat of Housing), joined forces to formulate a funding plan for the housing project. The solution chosen was to combine funding from the Ministério das Cidades (Ministry of Cities) with that from the state government to subsidize homes of social interest and recuperate buildings with families who had been living in the historic centre for 20 years by creating apartments of 26–55 square metres with one, two or three rooms.

In October 2007, the first renovated homes were handed over to the families. Of the 76 restored buildings, which resulted in 337 residential units and 55 commercial units, two buildings – one in *Rua 28 de Setembro* at number 10 and one in *Rua 3 de Maio* at number 21 – were able to house 11 AMACH families. Gecilda Melo, president of AMACH, said, 'This is a victory for our organisation: the apartments are excellent and appropriate for the family compositions and the residents are satisfied' (Plug Cultura 2007).

Conclusion

What took place in Salvador is an example of how resident participation in city management public policy can improve the overall quality of the interventions. For the first time in more than 30 years of politics to safeguard the historical centre, a programme that takes into account the citizens has been formulated. This represents at the same time a big success and a small step forward. Problems regarding the regeneration of Pelourinho are manifold and complex; it is necessary to think about an integrated programme that not only focuses on the restoration of the buildings, but that copes with the issues of economic and social sustainability.

Today, the Pelourinho looks like a true tourist district: a compact urban unit filled with characteristic squares and streets, noble residences and inner patios, where the most important and significant city churches and museums can be found. They are enlivened by restaurants, coffee shops, art galleries, souvenir shops, antique stores, book shops, jewellery stores and, above all, street performances. At the same time, there are five *favelas*, whose residents live in precarious economic and living conditions. Street vendors and prostitutes, drug dealers and artists fill the Pelourinho together with tourists, music and dance students, black suburban youths and a 'massive police presence, which is the guarantee of tranquillity that the market demands' (Santos 1995:28).

What remains now is to follow the project's progress in its successive phases. Will this small step build into a change in political approach, and lead to integrated and sustainable interventions while maintaining the cultural dynamics of the historic centre? Or will the seventh step remain an isolated event, and the Pelourinho be reduced to a showcase where tourists can visit without fear of coming into contact with the real Brazil, which is made up of contrasts – exclusion and poverty, and incredible strength and beauty that abounds everywhere.

18 Retail gentrification in Ciutat Vella, Barcelona

*Núria Pascual-Molinas and
Ramon Ribera-Fumaz*

In international urban commentaries, the 'Barcelona model' of regeneration is often held to be exemplary. It is said that the repositioning of the city to capture new flows of capital and create new public spaces represents both economic and social success (Marshall 2004; McNeill 2003; Sokoloff 1999). This repositioning came about largely through policies of the 1980s that were driven by the hosting of the 1992 Barcelona Olympic Games.

For some, however, the Barcelona model is not entirely progressive (see Garcia-Ramon and Albet 2000; Capel 2005). Its 'success' is based on increased gentrification and tourist flows in once humble but now fashionable quarters. Further, the pressure of maintaining Barcelona's status as a competitive city has brought new policies and strategies that are accelerating the regeneration of working-class districts and consolidated gentrification patterns. We have already seen in this book how these processes are expanding in Barcelona and facing strong opposition from neighbour and social movements (see Chapter 12). In this chapter, we will talk again about opposition and resistance to city council regeneration plans and the gentrification processes that followed their implementation. Yet, we want to focus our attention not on the struggles themselves, but on how they can shape new policy designs, and we analyse the potentiality of new policies that could contribute to build a city for the many and not simply the few (Amin *et al.* 2000). As an example, we will discuss two policy outcomes in Barcelona's Ciutat Vella district: a new garden and a new retail policy.

Ciutat Vella (Old Town), at the core of the city, contains most of the tourist attractions in Barcelona, and hence most of the tourist activity. It is also one of the most deprived areas in the city. Since the early 1990s, Ciutat Vella has faced a sustained process of gentrification, deeply affecting its residents' and neighbours' associations. Several policies have recently been put in place to attempt to ameliorate the negative outcomes of this process. We will discuss their effectiveness, and whether social struggle can shape the policy agenda towards a development strategy focused on the needs of the locals and not the attraction of foreign (wealthy) residents, visitors and capital.

Post-Olympic Barcelona model and Ciutat Vella

The post-Olympic scenario in Barcelona was shaped by the successful repositioning of Barcelona as one of the most admired and visited cities in Europe (Marshall 2004). This success supposed the return of local, national and foreign capital back into the city, making corporate interests a major force in pressing and shaping the city council's agenda. Furthermore, the early 1990s corresponded with the locking in of neoliberalism in Spanish politics through the application of the Maastricht Treaty and its stability pact. This, together with the financialization of the economy, the liberalization of labour markets, the privatization of public enterprises, and the return of centralized planning and spending centred in Madrid, have all added pressure for rethinking a new strategy for Barcelona.

The outcome of these processes has been an increasingly open neoliberal agenda to enhance the competitive position of Barcelona through supply-side economic policies, market competition, and the abandonment of redistributive concerns. In this context, neoliberalization in Barcelona has relied on new strategies and discourses that attempt to reposition Barcelona within the flows of the global economy. These are organized along three axes:

1   rescaling the city to the metropolitan scale;
2   positioning Barcelona as a knowledge, logistic and distribution, and tourist economy; and
3   maintaining the position of the 'Barcelona Brand' in the global competition among urban regions.

In this context, Barcelona's old town (Ciutat Vella: Figure 18.1) represents one of the most vital assets of the tourist strategy and its cultural and commercial supportive activities. Located right in the heart of the city, the medieval Ciutat Vella contains most of Barcelona's significant heritage landscape as well as new emblematic buildings such as Meier's Contemporary Art Museum. It also contains a lively commercial life, including both mainstream and alternative retail and catering facilities. Traditionally, the neighbourhood is home to a bohemian culture and leisure and a (not always legal) night scene. In this sense, Ciutat Vella has been a major contributor to the success of Barcelona, which

> has in part been based on its steady amassing of symbolic capital and its accumulating marks of distinction. In this the excavation of a distinctively Catalan history and tradition, the marketing of its strong artistic accomplishments and architectural heritage [...] and its distinctive marks of lifestyle and literary traditions, have loomed large, backed by a deluge of books, exhibitions, and cultural events that celebrate distinctiveness.
>
> (Harvey 2001:404)

But, surrounding the archipelago of regenerated and fashionable areas, there are derelict areas home to not only long-term poor and marginalized groups but also to new and quickly growing flows of migration. This combination of trendy areas next to a lot of devalued fixed capital together with a very dynamic real estate market in the city has also attracted young professionals and students from middle-class areas of the city and elsewhere looking for 'authentic' bohemian experiences. And of course, it has also brought contestation and struggles around the displacement of the poorest residents.

So far, this might sound the same story as many other places in North America or Western Europe. The particularity of the Barcelona case is the active role of the city council in facilitating the gentrification of Ciutat Vella since the arrival of democracy at the end of 1970s. Smith points out that 'third wave' gentrification is 'fuelled by a concerted and systematic partnership of public planning with public and private capital', and this is indeed what happened in Ciutat Vella right from the beginning (Smith 2002a:441; see also Hackworth and Smith 2001). To understand gentrification in Barcelona, we must start with the council-led regeneration of its central district.

Regenerating Ciutat Vella

Ciutat Vella is still one of the most deprived and densely populated areas in the city. The area corresponded to the actual city until the second half of the nineteenth century, when the medieval walls were torn down. The Old Town district physically retained some of the symbolic economic and cultural power institutions (e.g. the city council, the cathedral, the opera house) and several big interventions were carried out in the area when major thoroughfares were built (such as Via Laietana) and the 'Gothic neighbourhood' was constructed. But a gradual decline began. As Barcelona expanded following the modernist grid envisioned in the 'Cerdà Plan' (1859), most of the economic and political resources were concentrated in outward construction. The consequence for Ciutat Vella was a slow but deep process of physical and social degradation that lasted for over 140 years. The decline culminated in the 1970s, when problems produced by industrial restructuring resulted in a concentration of poverty, drug abuse and marginal activities.

Things only started to change with the decentralization of the municipal government and the first local democratic elections in 1979. A democratic left-oriented city council began to make plans to rehabilitate the area. During the first half of the 1980s, several Pla Especial de Reforma Interior (PERI – Special Plans for Interior Reform) focused on different areas of the district (Raval, Casc Antic and Barceloneta; see Figure 18.1). The first proposals for the PERIs were presented in 1982 and approved in 1985. The PERI consisted of 'surgical interventions' in specific spots (affecting around 15 per cent of the district), including expropriation of old dwellings for their demolition, eviction of their tenants, construction in the emptied space of public housing for (some of) the local residents, and 'sponged' reurbanization. In 1986, the whole of Ciutat Vella was designated an Area for Integral Rehabilitation and an ambitious Plan for Integral Action (PAI) was launched, complementing the PERIs with a masterplan approach. This included health and welfare infrastructure and several social-action and 'security and prevention' programmes. In 1988, a public–private owned company – 61 per cent city council, 39 per cent private – was created. This company, called Promoció Ciutat Vella S.A. (PROCIVESA; Ciutat Vella Promotion Ltd.) combined forced expropriations and reallocation of lands with profit seeking through buying and selling property to implement the PAI. Institut Català del Sòl built the new dwellings.

Figure 18.1 Areas in Cuitat Vella, Barcelona: 1) Raval Nord; 2) Raval Sud; 3) Santa
Caterina/Sant Pere; 4) La Ribera-Born.

Map by Núria Pascual-Molinas and Ramon Ribera-Fumaz.

During the 1980s, the interventions were limited to what were considered the
most urgent areas of the district. By the beginning of the 1990s, these interventions
had expanded, producing a division of the district into five areas: Raval Nord, La
Ribera-Born, Raval Sud and Santa Caterina/Sant Pere, Gòtic and Barceloneta.

Beside the good results remarked upon by the city council and by national and
international observers (press, architects) – improvement of living standards;

'cleansing' of drug dealers and prostitution; and the reinsertion of the district into the cultural, tourist and commercial circuits of the city – 2,000 residents were displaced by the Santa Caterina PERI, mainly to faraway areas, and 1,078 dwellings were demolished, including 13 buildings catalogued as architectural heritage (Mas and Verger 2004). In the Raval, 500 buildings were demolished from 1980 to 2002, 4,200 flats were lost and 2,725 new dwellings were built, of which only 1,245 were provided by the public administration (Subirats and Rius 2004). The process generated strong discontent and distrust towards the authorities, especially because of PROCIVESA's expropriations. Owners were offered very low prices and non-owners had few rights. The new buildings were built at low cost and are of dubious quality. There was no renovation of existing buildings, which remain dilapidated.

The cheap rents in the 1980s attracted young people to Ciutat Vella, especially students who couldn't afford to live in other districts. The 'social cleansing' prior to the 1992 Olympics and the 'rediscovery' of the area as a tourist attraction explain the gentrification of some parts of the district.

Yet, the newcomers are not all gentrifiers. During the second part of the 1990s, Ciutat Vella became the entry point for new economic migrants to the city. Foreign legal residents in Barcelona grew from 1.9 per cent of the total population of the city in 1996 to 15.6 per cent in 2007. Of Ciutat Vella's residents, 37.1 per cent are foreigners (Ajuntament de Barcelona 2007). Not surprisingly, they are attracted by the remaining affordable but insalubrious and almost obsolete housing stock. As a result of immigration and an accompanying process of renovation, there has been a sharp increase in the price of housing (see Table 18.1).

As the expropriations went on, insecurity increased, the social tissue of the area was eroded and much of the traditional commercial structure disappeared. Ciutat Vella's regeneration strategy was criticized for focusing only on quantitative

Table 18.1 Housing prices in Ciutat Vella, Barcelona (€/m²)

	1992	1996	2006	% change 1996–2006	2007
Price of newly built dwellings					
Barcelona	1,387	1,433	5,791	304	5,976
Ciutat vella	1,075	1,279	5,826	356	6,629
Ranking within 10 districts	8th	8th	4th		4th
Price of second-hand dwellings					
Barcelona	1,289	1,291	4,948	283	5,011
Ciutat vella	850	864	5,021	481	5,229
Ranking within 10 districts	10th	10th	5th		5th
Price of renting in second-hand dwellings*					
Barcelona	745	583	1,267	117	1,400
Ciutat vella	616	528	1,508	186	1,609
Ranking within 10 districts	2nd	3rd	1st		1st

Source: Ajuntament de Barcelona (2007)

Note: *€/useful m² per month

aspects (number of demolitions) and lacking a sociological or environmental analysis. The plan was rigid and difficult to change, particularly as the two administrations involved (local and regional) were governed by different political parties until the end of 2003. Members of the neighbours' association, critical architects and social movements in the area find it hard to understand that this approach has been exported to other districts in the city, and is cited as a model for other cities.

Struggles in Ciutat Vella for a change in policy direction

From parking lots to gardens

The dramatic transformations in the district raised some voices of protest, but perhaps not as many as could be expected given the magnitude of change and the long tradition of urban movements in Barcelona (Calavita and Ferrer 2000). Reasons for the relatively low level of dissent may be found in the levels of socio-economic and physical degradation in the district during 1980s and 1990s; the imagery of the area was stigmatized for many *Barcelonins*. At the same time, Barcelona was changing through the Olympics. After decades of abandonment by the local administration, the city council's marketing of the reforms used all the media available, ignoring the negative effects. But there was a persistent and incisive resistance to the council's plans. Despite the dismantling of social cohesion caused by displacement, demographic changes, and a general deactivation of neighbourhood movements in the city since the 1980s, neighbours nevertheless organized themselves on many occasions.

In the late 1990s and early 2000s, a number of new opposition platforms were created, such as Salvem Ciutat Vella (see Chapter 12), Coordinadora de Veïns del Casc Antic, Associació de Veïns en defensa de la Barcelona Vella, Col·lectiu de Veïns del Forat de la Vergonya, and Fòrum Veïnal de la Ribera. These involved actors from old urban social movements and new ones such as squatters, and parts of the new gentrifying population.

Emblematic of the resistance and of the struggle to force the administration to accomplish the plan in a way that was less damaging for the social fabric is the battle around what is known as the 'Hole of Shame'. Located in the Santa Caterina/Sant Pere area, the Hole of Shame was one of the sites cleared to make way for a green area in accordance with the original plan. After several years of abandonment (demolitions took place in 1999), the administration started a participatory process that involved citizen associations working with the council, but not the neighbourhood associations that were critical of the council vision. The outcome was a change of planning classification for the Hole of Shame from green open space to land devoted to neighbourhood infrastructure. It was decided that the necessary infrastructure was a car park for 150 cars for visitors to the neighbouring area of La Ribera-Born. Seeing the indifference of the authorities towards the renovation of their buildings, and the calamitous state of the demolished area, and the evictions of other residents, neighbours organized themselves with several demonstrations and protests. These culminated in an exercise of real

Figure 18.2 The 'Hole of Shame', Ciutat Vella, Barcelona. Self-made garden at the spot
left by demolition between streets Jaume Giralt and Metges, popularly known
as the 'Hole of Shame'.

Photograph by Marc Martí-Costa.

participatory democracy, as opposed to the one established by the district authorities. Neighbours discussed what was needed in the area and organized themselves to plant trees and build a children's playground (see Figure 18.2).

The city council's responses to the neighbours' actions were varied but the
garden and playground did not last long. In October 2002, as council workers
started to take soil samples prior to the construction of the car park, the self-made
garden was removed. The organized protests by the neighbours were then
repelled by a strong and violent police counter-offensive culminating in several
injured residents (one of them arrested) and a wall around the Hole. After this, the
presence of riot police on the site was a constant, but couldn't stop several popular demolitions of the wall. Eventually, the council reconsidered its plans and
declared that the Hole would become a green area.

In a wider context, neighbourhood mobilization and resistances such as this
meant that new policy approaches from the city council would be more sensitive
to the problems of local residents. In this sense, the area of Santa Caterina, just
around the corner from the Hole of Shame, a pilot strategy was undertaken. The
pilot attempted to solve another of the problems posed by the area's gentrification:
the disappearance of retail for low-income residents.

A new retail policy

One of the problems for residents of Ciutat Vella was the disappearance of retail providing for the needs of the poor and elderly (see Martin 2007). In the past decade, there have been many closures of retail shops, some of which were not replaced. Instead, the premises were often occupied by new fashionable music shops, designer outlets, bars and restaurants, directed largely at tourists and new bohemian residents. Only occasionally were they replaced by commerce and convenience stores that catered for the growing multi-ethnic population of the district and low-income locals.

The older residents, many of whom are low-income and with mobility constraints, faced big problems finding affordable places to shop. This issue was accentuated by the transformations of the key suppliers of affordable food in Ciutat Vella: the municipal markets. Following a policy drive concerned with modernization and health and safety, most of the municipal markets have been or are in the process of regeneration. Though this was necessary, their transformation is encouraging upmarket stalls in order to integrate the buildings with local tourist attractions. There is no clearer example of this policy than the regeneration of Santa Caterina's market (see www.mercatsantacaterina.net). This market is located near the deprived Hole of Shame and the tourist/gentrified area of Via Laietana/Plaça de la Catedral (Laietan Way/Cathedral Square). The regeneration of this area involved the redesign of Santa Caterina's market, incorporating a new roof inspired by Barcelona's nineteenth-century modernist style with a marked Mediterranean character, which fits very well in the tourist landscape. Inside the market, the space has been reformed. In the process, several small shops that didn't have the financial muscle to move to an alternative site during the reform process, and then pay more expensive rates for the new stalls, had to close, while a supermarket belonging to a nationwide retail group opened.

As a result of various other struggles between citizens and the city council, such as that around the Hole of Shame, the city council took a more nuanced and soft approach to the issue of retail services. A policy was introduced in which the council now buys closed premises and rents them at subsidized prices to businesses that will restore the commercial fabric and supply basic retail to locals.

Though originally set out in the 2004–2007 Ciutat Vella District's plan, the programme did not start until the end of 2007. It focuses on areas where there is less economic activity (outside the gentrified cores) with the object of bringing local social life back to the streets. Through a new public–private company, Foment de Ciutat Vella, the city council will pay an average of €2,500 per square metre for the closed premises, which is slightly under market price but reasonable for these areas. Once the council has a pool of premises, it will set a public competition to rent them, open to any sort of commercial facility and business. A committee of experts (including technicians, architects and retail associations) will evaluate all proposals, analysing which of them fit with the goals of 'preserving the commercial diversity of the area, maintaining local retail opportunities adjusted to the diverse socio-economic demography of the area, and

Table 18.2 Shops in Cambó, Barcelona

Name	Trade	Business type	# m^2
La Mallorquina	Home apparel	Franchise	187.39 + 166.61
Electrodomèstics Milar	Domestic appliances	Franchise	434.61
Jordi Manen	Architect office and pencil shop	Independent	75.91
Discos Castelló	Music venue	Local chain	71.58
MEPA'S	Fashion garment	Independent	121.01
Colmado del Zócalo	Bar	Independent	69.62
Andreu*	Delicatessen	Independent	206.29
Cotct	Optics	Franchise	
Castelló	Music	Local chain	
Benito Esports	Sport wear	Franchise	06.8
Icària Editorial	Publisher and bookshop	Independent	78.10 + 97.42
Epifanio Cano Guerra	Occupational school	Independent	127.30

Source: Foment de Ciutat Vella (2006a) and original research by Núria Pascual-Molinas and Ramon Ribera-Fumaz.

Note: *Split into three spaces.

guaranteeing the social dynamism of the neighbourhood' (Foment de Ciutat Vella 2006b). The committee will take into account the kind of activity planned and its economic viability, and develop a design for the chosen uses. The winning projects will be allowed to rent the facility for two years at a price below market, with an option to purchase it afterwards. If they take this option, they have to keep the business open for at least five years. Though aimed at non-commercial areas of the district, the process is in collaboration with and advised by the Ciutat Vella's retailer associations, which are organized along commercial streets.

At the time of writing, the programme has not been fully executed, but a pilot project has been implemented in the street between the Hole of Shame and Santa Caterina's market. As a part of the regeneration of the area, two social housing buildings have been constructed and their basements divided into 12 units that form the first pool of premises for the pilot. After an open competition where prospective tenants presented their bids, the expert committee selected the occupiers. The result is mixed in regard to the needs of the non-gentrified population. As Table 18.2 shows, there are new retail spaces that cater for low-medium market demand, such as the home apparel and domestic appliances franchises, and the occupational school. But some other shops are oriented more to the other end of the market, such as the delicatessen, or towards the new bohemian gentrified/tourist market such as the bookshop and the music venue. In sum, half of the new shops target the new bohemians of the district and the tourists that cross the area on their way to the gothic cathedral, the modernist Music Palace or the trendy area of La Ribera-Born, and the other half fit much better with the original idea of the policy.

While it is too soon to evaluate the results of the plan in terms of the rate of survival of the businesses, the effects on surrounding commercial tissue, the trickle-down effect on new commercial opportunities and so on, the intention of the council is to extend the policy to other targeted areas.

Policies to fight gentrification or gentrification as a policy?

Fights around the Hole of Shame ended up not only with the collective appropriation and redesign of that space, but led to other improvements such as the new retail policy. From the story explained here we can see the bottle as either half empty or half full.

If we take the first perspective, we can see how under a well-fabricated rhetoric and through practices that at first glance look progressive and respond to the effects of gentrification, the city council designs and implements a policy that contributes to support further processes of gentrification. Indeed, we can understand the process of Santa Caterina's regeneration as subsidizing capital to spur new investments and expand the gentrification archipelago in Ciutat Vella. Far from being an undesirable result, the expulsion of poor neighbours from the area through gentrification seems to be in perfect harmony with the plans of the authorities, easing the task of cleansing the area and reinserting it in the process of commercial accumulation. The disappearance of traditional retail or shops oriented to the needs of the non-gentry–locals accelerates their expulsion from their homes and makes the area more attractive for new waves of investment and middle-class in-migration. At the same time, new gentry retail aids the process of attracting the newly arrived residents and tourists. This is as true in Barcelona as it is elsewhere. To maintain, or not, a retail structure that caters for the working rather than the middle class is a vital issue in order to keep the social fabric of neighbourhoods at risk of gentrification (Zukin and Kosta 2004; Wilson *et al.* 2004). From this perspective, the 50 per cent of gentry-targeted premises show the real intentions of the city council, while the small victory on the Hole of Shame, reflected in the construction of a green area, may not be a victory at all but another factor in the area's gentrification.

But we can also see the bottle as half full. Neighbourhood resistance sometimes works and within the constraints imposed by local (and supra-local) neoliberal frames, city councils can think up imaginative, relatively inexpensive policies that do balance the effects of regeneration. The 50 per cent of the premises that are aimed at meeting the needs of poorer residents are, from this perspective, the first step of a change in city policy regarding the regeneration of the district, where meeting the needs of the local residents is a key objective.

As we have explained, the city council's strategy for Ciutat Vella must be positioned within the broader context of the efforts to place Barcelona in the neoliberal flows of globalization. It can be observed in the case of Barcelona, so far, that this is being achieved by attracting visitors, new gentry and capital to Ciutat Vella. Thus, negative change at the district level must also be fought at city and regional level.

Regeneration of long-term disadvantaged areas is a long process, as the Ciutat Vella story exemplifies. Its transformation started in the 1980s and it is far from finished. In this process, locals have fought, not always successfully, for a place for the many and not just the few. And if the process of regeneration is long, so are struggles for the right to the city.

19 The Melbourne indie music scene and the inner city blues

Kate Shaw

On a cold evening in May 2003, over a thousand people met at Trades Hall in Melbourne, Australia, to discuss the threats to the city's independent live music scene. For the Media, Arts and Entertainment Alliance, the Musicians Union and the organizers of the Fair Go 4 Live Music campaign, the cradle of the Australian union movement was a symbolic venue. At the meeting were musicians, performers, sound and light engineers, roadies, music writers, community media reporters and technicians, poster designers, printers and distributors, bar staff and others whose livelihoods depend on all the connecting art forms and supporting industries that constitute the alternative scene (Stahl 2004). They were clear about why they were there: inner city warehouse and factory conversions and in-fill developments were producing poorly insulated dwellings right up against old pubs and other venues, and the new residents from the suburbs (attracted by the *cachet* of 'inner city living') were complaining about the music *and* pushing up the rents.

Melbourne has been gentrifying unevenly for 40 years, and this was not the first rally against gentrification's effects. The focus on the music industry, though, and the implications of rising rents and venue closures for Melbourne's cultural scene in general, alarmed the governments that share responsibility for the Victorian capital. The State government and city council have long positioned Melbourne against Sydney as Australia's 'cultural capital' – without the corporate headquarters and 'global city' status, perhaps, but definitely with the cool. The council had already declared that 'a vibrant and independent arts sector, stimulating imagination and critical thinking about who we are, where we have been and where we're headed, is an essential ingredient in Council's vision for the future' (City of Melbourne 1999). Richard Florida's (2002) 'creative class' thesis had washed up on Australian shores, and John Brumby, the then State treasurer, was recommending it to his senior staff.

Here was a dilemma for the State and city governments. After a decade of explicit urban regeneration policies, the city brand was under threat. What to do?

The gentrification of inner Melbourne

The story of Melbourne's gentrification is a familiar one. Disinvestment in the post-war industrial inner city and its consequent 'undesirability' in the 1950s and

1960s for residential use meant that properties in the centre became run-down while the middle classes expanded into the new and growing suburbs. Houses and flats in the central municipality of Melbourne and surrounding districts (see Figure 19.1) were generally small, the rents were cheap and vacancy rates were high. Australia has a very high level of home ownership, at 70 per cent, so people who rented, through circumstances or choice, were already different to the norm. Hippies, artists, students, activists and drifters clustered in the inner city with the remaining working-class and immigrant populations, reversing the national ratio of renters to home owners, and making good use of the streets and public space. The empty factories and warehouses, and the old pubs that once serviced their workers, were perfect venues for Melbourne's flourishing alternative scene, which was developing a national reputation for its theatre, comedy and live music.

Melbourne's central business district (CBD) – made up largely of offices, shops and warehouses – had always had a very small residential population (in the mid 1980s, the official count was about 800 households [City of Melbourne 2007a]). Unoccupied buildings were informally converted to live/work studios by artists, designers and musicians with desire for large spaces, good light, and distance from their neighbours. In the 1970s and 1980s, this cluster came into its own after dark, when the office workers had returned home, with a cultural scene based on arts and music and experienced in small galleries and clubs.

Hackworth and Smith's schema of the three waves of gentrification fits Melbourne well. The first wave began in the late 1960s in a context of 'ameliorating urban decline' (2001:466) via government strategies such as restoration grants and easing access to home ownership (Logan 1985; Shaw 2000). The more elegant pockets of inner city terraces and workers' cottages gentrified quickly, but the vacancy rates remained high, the rents stayed low, and there was little if any displacement. The second wave in the 1980s was largely market driven and more extensive, and resistance to the loss of low-income housing and accessible public space fanned from Williamstown to St Kilda (see Figure 19.1). The lowest income renters were the first to be displaced. They were followed by the remaining working-class and immigrant home owners who capitalized on the huge gains to buy into the Great Australian Dream in the middle and outer suburbs. Small investment properties were also sold, setting off further rounds of displacement and reducing the number of places available for rent.

The global recession of the early 1990s ended the second wave of gentrification and the rapidly escalating property boom, and a decade of State Labor government. In 1992, an aggressively neoliberal government came into power on the promise of getting Victoria 'on the move'. A joint State government and City of Melbourne initiative called 'Postcode 3000' encouraged residential conversion of warehouses and offices in the CBD, led by model developments and financial incentives to building owners. Over 5,000 dwellings were added to the centre in the 1990s; since then this has doubled (City of Melbourne 2007a). A study of the CBD in 2004 found an 830 per cent increase in residents from 1992 to 2002 and, in the same period, a 275 per cent increase in cafes and restaurants (City of Melbourne and Jan Gehl Architects 2004). In the 15 years from 1991 to 2006, the

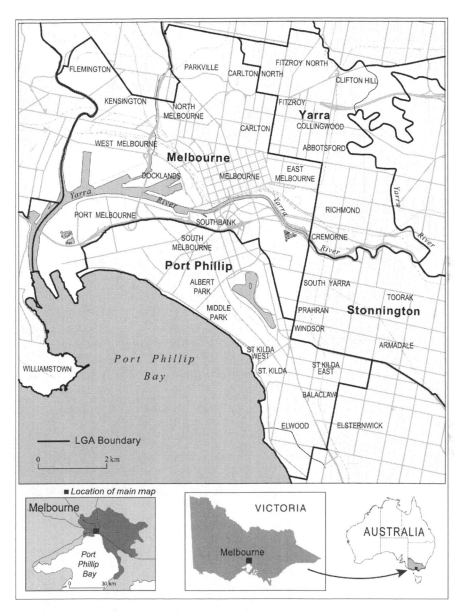

Figure 19.1 Map of the inner metropolitan region of Melbourne.

Map by Kate Shaw.

residential population of the whole municipality almost quadrupled, from 20,348 to 76,678 (ABS 2006). A major docklands redevelopment, begun in the late 1990s, had by 2008 produced 3,200 apartments with a median weekly rent double that of the Australian median (VicUrban 2008).

This third wave of gentrification was much more comprehensive than the first two, extending into areas that involved greater economic risk, requiring and receiving substantial State assistance (Hackworth and Smith 2001). Enabled by legislative amendments, special acts of parliament, planning and building deregulations and policies to encourage development and reduce public input, luxury residential high-rises, exhaustive factory and warehouse conversions, and medium-density in-fill housing spread from the city centre through the inner metropolitan region. A standard residential code was applied to land within a seven kilometre radius from the city centre, called the 'Good Design Guide', which produced a large amount of cheaply constructed but highly priced in-fill housing. Attacks on the union movement, which was already ailing with the loss of Melbourne's manufacturing base and transition to the service and 'knowledge' economy, complemented this strategy. Abolition of State awards, restrictions on rights to strike and to enter workplaces assisted the growth of a non-unionized, lower-wage labour force.

The uses of some of the buildings in the CBD occupied by artists were formalized, but many more were quietly removed. By the early 2000s, low-cost living and working space and adaptable 'public' spaces in the inner city were scarce. The working-class was long gone, and the small number of public housing units – provided by the State and originally intended for 'working families' – became increasingly targeted to the most disadvantaged. The Community Housing Federation of Victoria observed in 2004 that 'just being on a low income no longer ensures access to public housing. Most current tenants are in receipt of [welfare] payments, and are described as having "high" or "complex" needs' (cited in Commonwealth of Australia 2004:24). A reduction in real levels of government spending on public and community (social) housing was not met with an increase in low-cost private rental housing, leading to sharply increased rental costs at the cheaper end of the market (ibid.).

None of the newly built or converted dwellings in inner Melbourne were affordable to people on low incomes. The old dock sheds used for rave parties and sculpture studios and set construction were demolished. Many of the pubs were redeveloped, and the remaining music venues were under such pressure from new neighbours to turn the music down that they were losing their viability and being converted into bars more to the taste of the new population. The alternative scene was shifting to the outer suburbs and country towns, some of which were also showing signs of gentrification.

The peculiar pursuit of property in Australia

Australia has a very high number of small property investors: 17 per cent of taxpayers received rental income in 2004 (Commonwealth of Australia 2004). This is at least twice as high as in the United States and Canada, and about seven times that in the United Kingdom. The rental stock is therefore less stable than in societies with more institutional investors; indeed the small-scale buying, selling, demolishing, building and renovating of property is a vital component of the

Australian economy. A national Productivity Commission Inquiry report in 2004 found that 'over the past few years, an average of around 500,000 dwellings have been bought and sold each year ... in Melbourne, some 100,000 dwellings changed hands in 2001' (Commonwealth of Australia 2004:14). Bank lending for housing accounted for 65 per cent of loans to households at the beginning of the 1990s and is more than 85 per cent now. Much of the increase is due to the rise in the number of owner-occupiers who have acquired investment properties. Property investors account for one-third of banks' outstanding housing loans.

The situation is encouraged by federal policies that allow investors to claim tax deductions if their rental income is less than their borrowing and other costs. Further deductions can be made for 'capital works'. Sale of an owner-occupied home is exempt from capital gains tax, and individuals and trusts receive a discount of 50 per cent for investment housing held for longer than 12 months. These tax concessions are a form of middle-class welfare that gives profound advantage to private investors who, because they usually have a family residence, which serves as an additional or alternative form of collateral to secure loans, already have advantage over people who do not own property. This policy context is essential to the story that follows.

The effect of gentrification on the live music scene

By the early 2000s, important venues throughout the city were battling resident complaints and pressures for conversion to 'higher and better' uses. The famous blues and roots Continental Café in Prahran closed in 2001 after a dispute over the landlord's proposal to double the rent. Long-term grunge and rock 'n' roll venue the Punters' Club in Fitzroy closed in 2002 after the licensee and the landlord 'were unable to reach an agreement to renew the lease' (*The Age*, 11 October 2001). It reopened a little later as a bar called Bimbo Deluxe, with recorded music and drinks at twice the price. Fitzroy's Bullring, a Latin music venue, closed early in 2004 to make way for a proposed retail and apartment complex which would put a new block of apartments right behind Bar Open, another live music venue in Fitzroy. The Bullring's long-term owner summed up the situation most succinctly: he decided to close the venue because 'the land value was too great to run it as it was' (*The Age*, 19 May 2004). The Tote Hotel, an indie rock 'n' roll pub in Collingwood, was facing the probable residential conversion of its immediate neighbour, a recently closed Technical and Further Education campus. The Esplanade Hotel in St Kilda, one of Melbourne's most diverse music venues and long a cause célèbre, had been fighting off one redevelopment proposal after another and was at that time battling a 38-storey celebrity architect designed apartment block in its beer garden.

New dwellings next to live music venues often render compliant venues noncompliant with Environment Protection Authority (EPA) noise emission levels, as levels are measured not at the source but at the location where they give rise to the complaint – in some cases an open balcony. The publican of the heritage-protected Rainbow Hotel, a roots music pub in Fitzroy, says he was forced to

Figure 19.2 The Rainbow Hotel and its neighbours.

Photograph by Kate Shaw.

spend AUD$80,000 in sound-proofing (and fines and legal fees) to satisfy his new neighbours whose own levels of insulation were inadequate (see Figure 19.2). Other venues had to stop playing live music or go acoustic.

There was a growing feeling that when the remaining venues were redeveloped and gentrified, there would be nowhere else in the city to go. The national arts funding body, the Australia Council, commissioned an inquiry in 2002 into the state of popular live music and the resulting report, entitled 'Vanishing Acts', concluded thus:

> It is frequently the 'intruder' who demands that the rules be changed, and who has leverage because of increased council revenue (rates, land values, etc.). In attempting to resolve this tension it is by no means a given that the gentrification process should be privileged at the expense (literally) of the pre-existing local culture, and this extends to its relationship with existing music venues. This has relevance to local council attitudes to development applications. ... In proposing some recommendations we will suggest that music is a community resource and therefore a community responsibility.
>
> (Johnson and Homan 2003:43–4)

Melbourne City Council was in no doubt about the importance of the city's cultural scene to its position in the national and global economy. Its cultural policy stated,

The Council must develop its cultural role and promote its achievements in order to secure its national and international reputation as a city of innovation, cultural diversity and artistic excellence.

(City of Melbourne 1999)

In 1999, the neoliberal State government narrowly lost the election. A Labor government (with strong historic links to Victoria's trade union movement) was re-installed, and there was a sense of hope in Melbourne's alternative scene. Through advanced methods of information dissemination, astute use of independent and mainstream media, and strategically applied political pressure, participants in the scene organized.

Fair Go 4 Live Music and the Live Music Taskforce

The reasons for the pressures on live music venues were identified in the meeting at Trades Hall, and the State government was called upon to do something: invoke a principle of first-occupancy rights, require better sound-proofing in new buildings, anything to protect the music from the complaints of the new residents. The economic benefits to the city from the music industry and its associated activities were tallied, and the marketing and tourism opportunities of a 'vibrant cultural inner city scene' were pointed out. The media coverage was intense and partial: 'Live Music Lovers Losing to Residents' (*Sunday Age*, 27 April 2003); 'Is Australia's Live Music Scene Dying?' (ABC Online, 13 May 2003); 'The Creeping Suburbanisation of the Inner-City' (*The Age*, 26 May 2003); 'Inner-City Blues' (*The Age*, 11 June 2003).

The planning minister was quick to respond. She said: 'Melbourne's live music culture is internationally recognised and locally celebrated, but as more people move to the inner-city, tensions have risen about the noise coming from pubs and clubs' (Victorian Government 2003). Within two weeks of the Fair Go 4 Live Music meeting, the government announced a Live Music Taskforce to assess possible solutions. The brief for the taskforce was broad in terms of addressing the noise issues and resident complaints: it could consider building codes including sound-proofing, reassess the EPA's policies on noise, and warn potential property buyers that an apartment or home was near a licensed premise or entertainment venue (*The Age*, 6 June 2003). It did not include strategies to address development pressures for higher and better uses, such as ownership arrangements and other protections.

The taskforce included officers from the State planning department, the EPA, local government, housing and development industry associations, and representatives of Fair Go 4 Live Music. It discussed an increase in EPA music emission levels in designated entertainment precincts, raising building standards to require sound-proofing in new inner city residential developments, and various legislative and regulatory amendments to the planning system. The latter included emphasizing the cultural significance of the city's live music scene in the State planning system, where it would have statutory effect, and developing local planning policies and regulations to support the operation and maintenance of live music venues in given inner city areas (Live Music Taskforce 2003a).

There are precedents in the current planning system for such regulations. The new Melbourne Docklands are under a special zone created in the 1990s, which requires all residential developers to build in high levels of sound-proofing to prevent complaints about the (government-funded) centrepiece of the redevelopment, a football stadium. Another special zone requires similar standards in the central city (Victorian Government 2004a). The City of Melbourne's strategic planning document contains policy objectives for the CBD intended to limit the effect of complaints from residents:

> CBD residents make an important contribution to the 24-hour vitality and liveability of the CBD. However residential amenity in the CBD is not comparable to that of residential zones, and residential development must not compromise the CBD's other functions.
>
> (City of Melbourne 2002:86)

In St Kilda, the local council had already introduced a raft of amendments into its planning scheme which recognized the cultural significance of the Esplanade Hotel as an independent live music venue, and ensured that the use could continue by requiring the maintenance of its access, operational and service areas. In addition, the council had negotiated sound-proofing works above the legal requirement in the substantially revised, finally approved ten-storey residential development in the hotel grounds (for a more detailed account of this story, see Shaw 2005b). These events were observed by the taskforce members with interest. Six months after it first met, a 54-page report was released with much fanfare at the Rainbow Hotel.

The report covered an impressive range of options, and systematically presented arguments against the implementation of each. Only two policy interventions were recommended: to encourage music venue self-regulation through 'Environment Improvement Plans' (where managers negotiate mutually acceptable practices with the neighbours) and a planning practice note to implement the principle that 'the onus of responsibility for the cost of noise management ... should fall upon the agent of change' (Live Music Taskforce 2003b:41). No commitment was made to increasing the sound-proofing requirements for new developments next to music venues. No planning initiatives were recommended nor were the EPA noise emission levels raised, meaning that venues already in trouble were no better off.

The Rainbow Hotel was to trial the first Environment Improvement Plan. This essentially maintained the arrangement it was forced to introduce several years earlier when the new neighbours arrived – sealing off the main entrance when music is playing to create a sound lock, and requiring patrons to enter via a side-gate and through the rear garden. The hotel switched to acoustic acts during the week. Chick Ratten, the hotelier, said patronage had already declined by 30 per cent since making the changes (Ratten in interview, 2005). His lease was expiring in three years, and he said he wouldn't be renewing.

Given the seriousness of the government's reaction, why was the policy response so constrained? One reason can be found in the composition of the taskforce, which was so broadly representative that there was opposition within the group to virtually every initiative proposed. The EPA opposed the raising of acceptable noise levels in designated areas as 'too complex', the housing and development industry associations opposed additional sound-proofing measures as these would 'increase costs', the local council representatives were torn between the venues and the new middle-class residents. The government planners opposed new planning regulations on the grounds that they would add 'further complication to a planning system that was already seen by many as over-complicated' (Victorian Government 2004b, Appendix F:x).

The domination of government officers contributed to the conservative outcome. The State bureaucracy's resistance to change was compounded by the decision-making process, which the officer vested with implementation of the recommendations says was based on consensus. Hence the practice note to implement the onus principle – which was supported in the taskforce – faltered when it came to the details. The State planning officer explains that as no agreement could be reached, the practice note 'never quite eventuated' (personal communication, 2008).

A more fundamental reason, however, is encapsulated in a subsequent amendment to the State Planning Policy Framework Noise Abatement Policy (clause 15.05), which reads,

> Planning and responsible authorities should ensure that development is not prejudiced and community amenity is not reduced by noise emissions, using a range of building design, urban design and land use separation techniques as appropriate to the land use functions and character of the area (clause 15.05–2, amended May 2004).

The objective of reducing resident complaints about music venues is very clearly driven not by concern for existing venues, but by desire to ensure that new dwellings are minimally affected. The emphasis is firmly on minimal inhibition of development. The federally subsidised and State-supported buying, selling, demolishing, building and renovating of property is simply too important a part of the city's economy.

> Third-wave gentrification has evolved into a vehicle for transforming whole areas into new landscape complexes that pioneer a comprehensive class-inflected urban remake. … Most crucially, real-estate development becomes a centrepiece of the city's *productive* economy, an end in itself, justified by appeals to jobs, taxes, and tourism.
>
> (Smith 2002a:443, original emphasis)

'Creative city' or sound cultural policy?

In December 2004, the Victorian State government brought Richard Florida to Melbourne to hear him say, 'I think it's obvious what you have done here is truly amazing'. Treasurer John Brumby replied: 'I think there's a lot of truth in his thesis. Melbourne itself … is proof of it' (cited in *The Age*, 4 December 2004). At the time, the first re-sales of Docklands apartments were not reaching their original sales prices (Real Estate Institute of Victoria 2004) and agents were warning potential investors in inner Melbourne that 'inadequate services and infrastructure and poor acoustic privacy because of cheap construction [are] significant drawbacks associated with renting in high-rise towers' (Wakelin 2004:3). The State economy was slowing. Inner city apartments were in surplus, building activity had dropped and rents had stabilized.

The economic downturn is the old ally of alternative cultural producers – not of Florida's 'creative class' necessarily (which consists essentially of anyone with a university education and white-collar income) but of those who experiment and take risks and often live on very low incomes. Development pressures on venues eased, and resident complaints let up too: the Fair Go 4 Live Music campaign and the weight given to the issue by the formation, at least, of the government taskforce, perhaps gave inner city residents pause before demanding their rights to peace and quiet. The Bullring sat empty and undeveloped; Bar Open continued its live music, theatre and comedy performances. The closed campus next to The Tote was converted to a neighbourhood justice centre with community meeting facilities that were used only during the day. The Tote played on.

The significance of the city's live music scene was recognized in a policy of the State Department for the Arts. Melbourne City Council revised its cultural policy to emphasize the 'venues, streets, laneways, buildings and parks [that] provide a public domain where art can happen and people can participate and engage' (City of Melbourne 2004:14). For the first time, it contained a direction to 'implement mechanisms to ensure artists live, develop and present their work in the city' (ibid.). A review of the policy in 2007 gave particular attention to 'housing the arts' in the city and included recommendations to further pursue low-cost artist housing (City of Melbourne 2007b:3). The council encouraged new, small bars and music venues in its laneways, and celebrated the stencil art and graffiti in the city's hidden spaces.

Cities never sit still. When a privately funded 'refurbishment' of one of these laneways was proposed late in 2007, involving the conversion of the most celebrated of the new small bars to the entrance to a 'very Melbourne boutique shopping idea', it was greeted by the council's planning chairwoman as 'fantastic news for Melbourne' (Lucas and Collins 2007:5). By 2008, the oversupply of apartments was fully absorbed. Rental vacancy rates in Melbourne fell to their lowest ever, below 1 per cent. Housing affordability across the nation is at an historic low, and building approvals are increasing again. The Tote has been sold to the hotel chain that turned the Punters' Club into 'Bimbo Deluxe'. The lease expires in 2010 and the new owners are not being drawn on their plans. Chick Ratten gave up his lease on the Rainbow at the end of 2007, as he said he would, but a month later the lights were back on with a musical line-up not far from the original, though quieter and finishing earlier. The new hoteliers, young but with experience, say they believe they can manage the live music venue and the relations with the neighbours. They are unencumbered: it just may be that with a conciliatory approach and good neighbourly skills, they can make it work. They are on a short-term lease: the owner will reassess the rent if they prove their success, or perhaps weigh in with a residential conversion when the time is right.

The different levels of government in Melbourne, and different departments within these governments, have different and competing objectives. There are clear tensions, and the ways these play out are equally clear: the 'cultural vitality' of the city will be supported to the extent that it does not impede property development. But Melbourne's 'cool' has as much economic as it does cultural value, and for every development advocate in the city planning department, there is a policy-maker in the arts and culture branch reminding the council of the importance of the city's recognition on the cultural 'world stage'. There is no resolution to this tension: contradictory and compensating policies will continue to dance around the line between the valuable music scene and the priceless property market. More productively, the exposure of these contradictions and the presence of potential and publicly aired policy interventions – to provide and support low-cost housing and performance spaces, for example – encourage dissent and debate. Melbourne's 'renaissance' remains hotly contested.

What is not in contest is that the city has changed profoundly in the past 40 years: not only is the working-class long gone, but there are fewer places for 'uncreative' low-income earners, too. The people who resist the city's gentrification and argue for more affordable housing are also those who oppose the loss of the city's cultural base, recognizing that in at least one sense, cultural producers now occupy the place of the old manufacturing workers. The mighty Trades Hall building, which is still used for Trades Hall Council meetings, though these days these are small affairs, is again a powerful symbol of the times. Its main use, as a low-cost independent theatre, comedy and live music venue, is an irony its custodians understand only too well.

In memory of Chick Ratten.

20 The embrace of Amsterdam's creative breeding ground

Bas van de Geyn and Jaap Draaisma

From 2000 on, the municipality of Amsterdam has actively subsidized local sub-cultures by means of a *broedplaatsenbeleid* (BPP – breeding place policy). This policy preserves cultural breeding grounds by investing in property for subcultural groups. Amsterdam had a thriving subculture in the 1970s and 1980s, with squatted buildings providing the basic infrastructure (Davies 1999; Soja 2000). The gentrification of the city at the end of the 1990s saw these squatted subcultural places disappear. In 1999, the local government came to the realization that safeguarding subcultural breeding grounds was essential to the contemporary space economy. Under the name Breeding Places Amsterdam (BPA) the local government began to subsidize buildings for artists and other subcultural groups.

In this chapter, we analyse the origins and effects of the BPP on subcultural Amsterdam from an insider perspective. In 2006, the BPP faced huge budget cuts. A green social democratic coalition came to power in the city government with a new but neoliberal programme called Amsterdam Topstad (Topcity). The green and red parties are following the (neoliberal) advice of Florida (2002), marking the change to the 'powerful city' where investment in urban qualities, with economy and culture the main gateways (Lupi 2007), replaces endless social-economic subsidies. The support of subculture shifted from safeguarding the market to focusing on its potential to attract investors. As a result of budget cuts, the local municipality has been outsourcing the BPP to housing corporations, real estate developers and users' group organizations. Subcultural groups are placed in an ambiguous position, for as culture becomes a valuable asset in the upgrading of urban space, artists and activist groups became part of this process. What position do these new subcultural landmarks have in the contemporary neoliberal urban arena? Have these subsidized subcultural places become catalysts for urban regeneration? And to what extent are these places now subject to processes of gentrification?

My creativity as a revolutionary tool

For a better understanding of the emergence of the BPP, we must look back in history. Amsterdam has been known for its authority-undermining subculture from the sixties with the Provos – the 1960s provocative resistance movement that shocked Dutch society by radical manifests and playful actions in public space, in which authorities played most of the time the leading part. Roel van Duin, one of the founders of the Provos, described the role of creativity in society as follows:

> my life should in the first place be aimed at the maximal exploitation of my creative abilities. ... the creative men will break as deeply as possible with the outside world and the society in order to become loose and unburdened ... that means for him: revolutionairism [!], not as a tool to obtain better living conditions, but as a goal, a creative power.
>
> (Van Duin, cited in Verschueren 2003:134)

Only your creative power can make you free from society, which is the only way to be really revolutionary.

When the squatting scene emerged in the 1970s, it had a highly pragmatic approach and one political goal: expropriation. But soon it became clear that living in a squatted building meant more than just a roof over your head; it was a way of life. From the mid-seventies on, an autonomous scene emerged out of the squats, with their own radio and television broadcasting stations, their own newspapers, music, restaurants, bars and jobs. The adage was 'do it yourself' (DIY), be autonomous from the existing order and create your own order. By being creative it was possible to change your life and live the life you wanted. The squatted building was the 'freespace' which made this life possible (Breek and de Graad 2001:16–17). In this respect, the squatting scene can be directly linked to the Provos in the sixties, and in the freespaces, creativity was seen as a way to becoming completely autonomous.

Urban guerrilla: from housing shortage to freespaces

At the end of the 1970s, there was a huge housing shortage and at the same time, many unoccupied dwellings. The average house price dropped from 198,800 guilders (about 90,000 euros) in 1978 to 138,100 (about 65,000 euros) in 1982, and housing speculators left their properties empty. The demolition of old dwellings exceeded the production of new social housing (Boelhouwer 2002). The city planners began to prepare huge regeneration plans to bring an end to the 'deprived' nineteenth-century workers' neighbourhoods.

Resistance to these plans was strong, and residents of these 'deprived' neighbourhoods took matters in their own hands. They founded their own neighbourhood centres and community organizations, and an autonomously organized housing distribution programme. These local actions coincided with the goals of the squatter movement in expropriating housing speculators, instigating direct action and creating a DIY economy based on self-activation (Draaisma and Hoogstraten 1983). The squatters were acting against the housing shortage and the destruction of the social and physical fabric of these neighbourhoods, ensuring them much support from locals (Duivenvoorden 2000).

The squatters' movement was a significant political force, partly because of the widespread support of local inhabitants but also because of their radical approach. The year 1980 will always be remembered in Dutch history by the picture of army tanks rolling through the streets of Amsterdam. Thousands of squatters barricaded central roads for days in order to resist the eviction from a squatted house. This forced the local authorities to call in the help of the army. At that time, the squatters' movement in Amsterdam counted around 9,000 squatters (Duivenvoorden 2000).

In the aftermath of the 1980 riots, the national authorities developed a Keynesian policy to deal with the housing crisis. The national authorities would invest in social housing in order to stimulate housing production and to deal with the housing shortage – a policy that was to influence Dutch housing production until the late 1980s (Kempen and Priemus 2002). Part of the policy included an indirect measure to legalize squats. Housing cooperatives, which in those times were public organizations, could buy squats from their owners and turn them into youth housing (Duivenvoorden 2000).

Between 1978 and 1985, around 140 housing blocks were legalized via this means, comprising thousands of living units. After legalization, the former squatters had a right to return. This legalization policy was acceptable to the squatters because the expropriation of ownership was guaranteed, and the housing was embedded in a housing cooperative that was not allowed to sell the property. But at the same time, the legalization of squats for youth housing can be seen as a step towards the institutionalization of the squatter movement (Pruijt 2003). In the legalized squats, the once variously used spaces were converted to single-use living spaces. The impact on the character and use of the buildings was huge, and the danger and fear of losing their way of life became as urgent an issue for some squatters as the fight against the housing shortage (Duivenvoorden 2000). The fight for housing evolved into a fight for free space within the boundaries of the legalization policy.

In the beginning of the 1990s, preceding the 'resurgence' of Amsterdam, the last wave of legalized squats took place. By then the economy had recovered and the housing shortage was less urgent. There was less pressure to build more social housing, and the legalizations gave more freedom to the squatters. Between 1990 and 1994, around 40 buildings containing around 1,000 units were legalized. First, the estate company of the municipality handed over all its squatted property to the squatters at symbolic prices. (This action was done under the euphemistic name, *schoonschip* [clean ship]). Within these new 'owner-occupant' circumstances, the squatters could decide how to reconstruct their places for themselves. Then a group of buildings, the so-called Casco-buildings, which was already owned by a housing cooperative, was rented to a collective of squatters. The housing co-op took responsibility for maintaining the outside of the buildings; the inside was controlled by the inhabitants. By means of both these measures, it became possible to create self-managed buildings and realize the concept of mixed-use (Breek and de Graad 2001). Perhaps the legalizations derived from the city council's lack of interest in the buildings and greater interest in neutralizing Amsterdam's last anarchistic bulwarks. But these measures can also be seen as heralding a growing interest from market players in freespaces, in which more space is provided for experimental ways of living, and more 'ownership' for renters.

Urban renaissance: where have the freespaces gone?

In the mid 1990s, the urban landscape of inner Amsterdam changed rapidly. The city had entered the information age, with the national airport Schiphol the gateway to its globalization, along with key sectors like banking and financial services. The inner city canal structure and old workers' neighbourhoods like the Pijp and Oud-West became neighbourhoods for the new urban middle-class (see Figure 20.1). This was the heyday of Dutch neoliberal 'third way' politics, with the former socialists at the helm for the best part of that decade. One of their most influential achievements in urban space was the liberalization of the housing cooperatives. Social housing became the responsibility of privatized housing corporations which were no longer accountable to local government. In order to become financially sustainable, these corporations had to sell dwellings. In regenerating areas, the amount of social housing dropped by an average of 20 per cent (Kempen and Priemus 2001).

The renaissance of Amsterdam led to two developments for the subcultural scene. First, the squatters turned out to be forerunners for the renaissance of the city. The needs of the 20-something generation in particular have been seen as important to the economic progress of many cities (Soja 2000). The origin of Amsterdam's vibrant clubbing scene lay in the squatter's movement. Established pop temples in the 1990s started up as squats in the 1980s. A former church, the famous club Paradiso was squatted in 1968 and was a venue for punk bands, radical expositions and autonomous publishers (Duivenvoorden 2000). Now it is the biggest pop temple of the Netherlands, providing a legendary performance stage for pop bands. Out of the squatting movement, an autonomous media network emerged. It provided a framework which could be termed 'sovereign media, a network liberated from the statutory obligation to serve an audience and benefit shareholders' (Adilkno 1998, cited in Riemens and Lovink 2002:327). By the end of the 1990s, some of these initiatives had become major IT players.

A second consequence of Amsterdam's urban renaissance was that freespaces became limited. This meant an end to many landmark subcultural squats. The only remaining freespaces were in warehouses in deserted harbour areas. Until the late 1990s, these warehouses, some of which, like Vrieshuis Amerika and the Graansilo, had become famous, provided space for an experimental subcultural scene from skateboarding halls to performance theatre. The first plans for restructuring these harbour areas were prepared in the beginning of the nineties; by the end of the decade they had become reality. Nearly all the squatted warehouses were evicted to make way for new waterfront developments and lofts for the new gentry. Amsterdam was booming, and the regeneration of its last unoccupied spaces meant a definite victory for unification and a step towards 'boorification' (Uitermark 2004:238).

No culture without subculture: towards a breeding place policy

The eviction of these landmark squats led to huge protests. When in 1998 a group of activists wrote a council address in the aftermath of the eviction of the Graansilo, to their own surprise the council responded. The activists had asked for recognition of subcultural landmarks and for their protection from future urban regeneration plans. The council established a working group, which recommended investing 100 million guilders (40 million euros) in subcultural infrastructure. Paradoxically, this investment was possible because of the huge increase in the prices of the council's own property holdings.

The motivation to subsidize subculture can be understood through the importance of culture in the contemporary space economy. In this respect, one could speak of a new cultural competition with cities like Brussels, Berlin and Rotterdam. Amsterdam had a disadvantage; in this city, it was no longer possible to rent working spaces for low rent. On the cultural level, Amsterdam was losing its innovative image to Rotterdam. In 2001, Rotterdam was elected European Cultural Capital. In the years leading to this, investment in culture had become a major interest of the traditional harbour city, and Rotterdam had already attracted influential cultural institutions from Amsterdam, including the National Museum for Photography and the Berlage Academy for Architecture.

For Amsterdam City Council, it was important to regain some cultural momentum. In the year 2000, the city introduced the BPP. According to the municipality, 'if small experimental theatres, cinemas or workspaces are swept out of the city it is not possible to have a culturally strong city' (PMB Amsterdam 2000:1, translated by authors). The official slogan of the municipality became 'No culture without subculture' (PMB Amsterdam 2000:1). This meant a fundamental change. In the 1970s and 1980s, subcultural and pop cultures were mainly regarded as rebellious youth cultures, as 'the dirt and disorder' of society (Kamsma 1998:112). From the time of the BPA, subculture was seen as a cultural breeding ground in need of protection from the market.

It was recognized that some types of artistic and subcultural activities could only exist when a sufficiently large supply of relatively cheap space was available. To counterbalance the spiralling land rents and prevent resulting processes of exclusion, the local government had to intervene in the housing and land market by means of a financial injection to guarantee the availability of such space. The BPA was a policy to provide subcultural breeding grounds, and the municipal department of Housing and Urban Planning was responsible for its implementation. The BPA had a budget of 41 million euros for the production of between 1,400 and 2,000 workspaces for two groups: individual artists and user group collectives. The demand of around 1,500 workspaces was to be met over a period of five years (PMB Amsterdam 2000:1).

Characterizing your own freespace

It took one more eviction before the BPP would really be installed. A year after the eviction of the Graansilo, the last squatted subcultural landmark, Kalenderpanden,

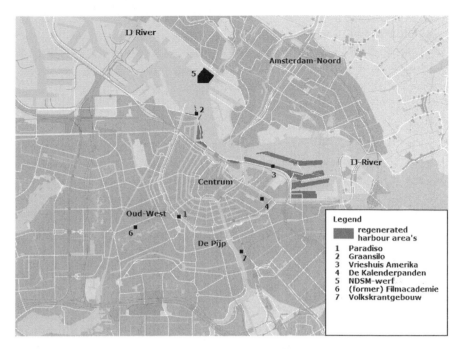

Figure 20.1 Map of Amsterdam and key creative sites.

Source: © 2008 Gemeente Amsterdam, Geo en Vastgoedinformatie.

came under threat of eviction. A widely supported coalition, newly established cultural institutions standing back-to-back with high cultural institutions, opposed the eviction. Ensuing riots meant that Kalenderpanden became the most violent eviction in Amsterdam in years (Uitermark 2004:234).

After the eviction, a number of former squatters came together to create an action group called *de Vrije Ruimte* (freespace). This group played an active role in the implementation of the BPA. One of its achievements was to shape the policy to include not only artist spaces, but spaces for craftsmen, activists and other creative producers. In September 2001, de Vrije Ruimte published a study on freespaces in Amsterdam: *laat 1000 vrijplaatsen bloeien* (Breek and de Graad 2001). The group articulated the characteristics of a freespace. First of all, the buildings were obliged to allocate space to mixed-use; there had to be a combination of 'living and working' spaces in the former factories and schools. Second, a DIY, or self-management principle was applied. The organization had to be autonomous, for example, the selection of newcomers was to be done by the users, not by the municipal housing distribution system. Third, the organization had to take part in a collective: every user was to have a vote without any form of hierarchy. The last characteristic that defined freespace was that the rent had to be cheap and the place should be open to everybody. In this regard, the provision of public functions was essential, for example, a theatre, nursery or political centre (Breek and de Graad

2001:16–17). The most important difference between freespaces in the seventies and today is that the public function of contemporary freespaces is for the city and is not there only to secure the autonomy of the squatter's movement. The characteristics formulated by de Vrije Ruimte are largely incorporated into the BPA.

By looking at two examples of breeding spaces, we will now give some insights into the way the BPA has functioned since 2000.

NDSM-Werf

Nearly half of the planned square metres (20,000 m^2) of the BPA were realized at the NDSM-Werf (NDSM-Wharf). This former shipping wharf, on the north side of the IJ-river, is part of an urban regeneration plan. The re-use of the wharf was initiated by the local borough Amsterdam-Noord. Tenders were invited from groups that could transform the production hall into a *Kunststad* (Art City). A group of squatters under the name *Kinetisch Noord* won the competition. In the end, the project was a co-investment involving the borough of Amsterdam-Noord (4 million euros), the state (1.5 million euros), and the BPA (7 million euros). This money was spent in the reconstruction of the shell of the hall. In 2007, the project was realized and the Kunststad now consists of 200 built-in units built by the artists and craftsmen themselves. It includes a skateboard track and offers free-space to outside users for theatre performances, dance clubs, large festivals and massive techno parties.

Despite the expectations of the users, the subcultural Kunststad has rapidly attracted market players. MTV (Belgium and Netherlands) opened its new head office in the area, the Dutch media company ID&TV settled on the wharf and the first new urban chic restaurant opened in its grounds. The still unaffected industrial urban landscape of the wharf and the vicinity of Amsterdam's street culture were decisive factors in MTV's decision to locate there (*Het Parool*, 15 November 2005). MTV has committed to actively taking part in the development of the shipping wharf, promising to develop a new street culture hall and invest in the management of public space.

To a large segment of the users group, MTV's settlement made them feel that they had been used to kick-start the regeneration of the terrain. Their sacrifices had been huge: rents had doubled from the original agreement (from € 25 to around € 50 per square metre per year). The need to comply with safety regulations has seen their costs rise, and delayed the project for years. The users had to comply with the selection criteria of the BPA (management skills and financial and administrative reliability) which made breeding places inaccessible for 'non-professional' artists and craftsmen. At the end of 2005, the users of Kinetisch Noord wrote a provocative letter to the BPA entitled 'The end of the breeding place policy'. According to them, 'artists have become the prostitutes of Amsterdam and the BPA is their brothel owner' (*Het Parool*, 15 November 2005). The irony of obtaining their freespace is that they had helped change the surrounding landscape into a place of which they never wanted to be a part. The users' appropriation of the breeding places seems to be at stake in the regeneration of the NDSM-Wharf.

The former Filmacademie

Under Dutch regulations, squatters who occupy a building within a year of its last use can be evicted. The former Filmacademie was empty for only a couple of weeks, but squatters convinced the BPA and the local borough of the importance of this space as a breeding place for young artists. Since 2000, this breeding place has been the centre of Amsterdam subculture, with an active experimental film programme, theatre performances and political lectures. The Filmacademie took the position within Amsterdam's cultural infrastructure of some of the earlier subcultural landmarks. A flaw of the BPA is its limited ability to allocate space for public use in the breeding places. From the start, the policy was mainly focused on creating alternative working spaces for the user groups. In this regard, the breeding space policy has been successful. The user group in the Filmacademie has managed to maintain a subversive subcultural character within the institutional frame of the BPA.

But public activities were prohibited for nearly one year because the building did not comply with contemporary safety regulations. As the main activity of the group was the organization of public activities, this did place pressure on the future of the Filmacademie. The users had to fight for their public function which nearly led to their end. Paradoxically, in 2007 the Filmacademie has been rewarded with the most important cultural prize of the city of Amsterdam. This contribution of a breeding place became an important asset in the next phase of the embrace of Amsterdam's subculture.

Topstad Amsterdam: from culture to creativity

In 2005, the first phase of the BPP ended. The Amsterdam economic boom had slowed and Amsterdam seemed ready for a new injection. In 2006, a green–red municipal coalition came into power. Their economic programme takes the next step in third way politics. It meant a definite end to the traditional social democratic task of reducing the negative effects of capitalism and compensating those who suffered most from the globalizing market. According to the coalition, in a globalizing economy the only way to provide opportunities for *all* was to invest in a powerful city. The economic policy was summarized in a report with the ambitious title 'Amsterdam Topstad'. The central point of this report was that Amsterdam should be brought back into the top five of Europe's locations for international companies. Amsterdam had to 'change from a city into a metropolis' and creativity should become the motor of the new economy and distinguish Amsterdam from its competing European cities. Culture was recognized as 'essential' to the establishment of a 'creative environment' (College Amsterdam 2006:2).

From then on, the subculture debate was placed in a completely different perspective. The BPA continued to exist, but its budget was cut to a direct investment of 1.4 million euros a year. In terms of appropriations, the BPA no longer took the lead. Coalitions were required and housing corporations, real estate agents and user groups were asked to take the initiative. Topstad Amsterdam, part of the economic department of the municipality, became the leader of council investments in creative breeding places, with a budget of 51 million euros over four years. The case for experimentation was no longer enough to warrant investment: private partners had to be willing to invest in a project.

The argument for financing subcultural activities to protect them from the market was now completely reversed, and investment in cultural activities was an explicit catalyst for market-led regeneration. The BPA now looks for public–private partnerships with user groups as well, forcing applicants to attract money from external partners. In this new context, the freedom to take the initiative yourself means that only established artist groups can start up working spaces themselves.

Do it yourself: the former Volkskrant building

By the beginning of 2006, over-investment in office space in Amsterdam had resulted in 1.4 million square metres of unoccupied office space (18 per cent of the total amount (O+S 2007). A group of activists around de Vrije Ruimte formed a foundation called Urban Resort, with the object of taking advantage of the situation. In June 2007, Urban Resort, with BPA support, rented a 10,000 square metre office building owned by the housing corporation Het Oosten.

The building was the former office of the second largest newspaper in the Netherlands, *De Volkskrant*, a workers' newspaper, and was part of a large regeneration programme. The housing corporation agreed to rent the Volkskrantgebouw to Urban Resort for five years at low rent and under reasonable conditions. A coalition of partners – the housing corporation, Topstad Amsterdam and the BPA – would finance the necessary investment in the written-off technical installations. For the city council, the building would become a temporary cultural hotspot. For the housing corporation, a cultural hotspot at the centre of its property holding meant an attractive creative environment to attract the new middle-class.

In just a couple of weeks, Urban Resort opened the building. The foundation organized the building according to the characteristics of legalized squats as defined by de Vrije Ruimte. Affordable rent and accessibility for all users became possible by means of the ability-to-pay principle, with prices around that of the lowest social housing rent. The highest commercial rent in the building was set at just below the average commercial rent in Amsterdam. Self-management was a necessity for cost reduction, and also in order to stimulate appropriation by the renters. Urban Resort would control the basic services such as heating, ventilation, Internet connections, water and electricity, and users were to manage the rest of the building (reconstructions, cleaning, security) by self-management.

After the first informal call for tenants, it became clear that the demand for cheap workspace was enormous (as we write there are around 2,000 applicants for 270 spaces). In spite of eight years of the BPP, the demand for cheap living and working spaces is growing in gentrifying Amsterdam. To meet the requirement for self-management, people have to apply in groups or collectives. Selection criteria were formulated based on what the applicants' thoughts were about self-management, their common interests, and what their contribution would be to the building, the neighbourhood and the urban perspective. There are no individual selection criteria.

The renters in the Volkskrantgebouw are very different from renters of other breeding places. They range from open source ICT businesses, fashion designers, woodworkers, photographers, theatre producers, animators, ethnic broadcasters and development organizations. Two collectives are non-white groups, and almost 40 per cent of the building is rented to non-white people: an exceptional phenomenon in Amsterdam and the Netherlands where the subcultural scene is predominantly white. To create public spaces in the building and to open the building to outsiders, the building is rented to entrepreneurs: a youth organization, a dance studio and a restaurant/club. Four public halls are rented at moderate prices to renters and outside initiatives. The public function of the building remains uncertain, however, as the building has to comply to ever-strengthening building and safety regulations.

The position of Urban Resort is, at first sight, a continuation of the squatters' activities over the last decade, creating affordable subcultural spaces for artists, craftspeople and activists. As a developer of an office building in the centre of Amsterdam, Urban Resort attracts new user groups to subcultural spaces. For migrant groups from the outskirts of the city, it offers the possibility of they too appropriating part of Amsterdam's urban culture. At the same time, the organizers and former squatters have had to meet certain state-sanctioned criteria. These criteria were originally part of the political fight to move urban space towards freespaces, but have now become institutionalized. They were formulated in order to continue the activities of a subversive squat, but are now the criteria for entering a 'cultural hotspot'. At a temporary level, it has been possible to safeguard the space against gentrification, but the Volkskrant building has also become part of the regeneration process the squatters had tried to oppose.

When looking back on the last ten years of the BPP, it becomes noticeable that there is no space for new subcultural squats. In 2008, parts of the Red Light district are being closed down by the local authorities to make legal space for artists and other creatives. Within the gentrifying frame of Amsterdam's social control, uniformity and the creative city merge, and space for low-rent housing, diversity and tolerance is under threat. According to Žižek (1998), the Provos had it right when they claimed that creativity is the most subversive power to break with the surrounding society. In the contemporary neoliberal society, creativity has become its most valuable mode of production (Žižek 1998). The position of 'subcultural' in Amsterdam has completely altered: in the 1960s, subculture was the 'dirt and disorder' of society, but now, in the 'creative city' a crazy and exciting subculture is embraced.

21 The equitable regeneration of Berne

Angela Stienen and Daniel Blumer

This is a story of contested gentrification in a small inner city neighbourhood in Berne, the capital of Switzerland. The district is called Lorraine, and it is a case of urban regeneration that did not displace any residents. Berne is often considered Europe's 'most unimportant capital', as the Swiss newspaper *Facts* once ironically put it (Friedli *et al.* 1997). Time seems to have frozen in the city's picturesque historic centre, founded in 1291. The medieval facade resists change, especially since Berne is classified as a world heritage site by UNESCO. Berne is ordinarily seen as a placid city, dominated by protestant reserve and a bureaucratic mentality that is antagonistic to change. Nevertheless, Berne seems to tolerate beggars and has a liberal drug policy. This at least is the impression you have when crossing the historic centre: under Berne's typical arcades again and again you walk past figures who cower on the ground, in front of them a tin and a lettered card asking for money. Often, you are also kindly requested to give money by respectably dressed people, young and old. On Sundays, in front of Berne's central station, you could encounter the same performance over and over again: young people belonging to an anarchist-inspired college association are publicly catering for drug-addicts with hot soup and tea. These forms of public space appropriation are tolerated by Berne's red–green coalition which has governed the city since 1992, but they have always been opposed by the city's conservative and right-wing parties. These parties want to ban beggars and 'clean' the historic centre of drug-addicts and the city's undesirable 'others' as is happening in other big Swiss cities. How do such contradictions shape everyday life in Berne's inner areas as they undergo the kind of regeneration processes undergone by inner cities all over the world? Using Lorraine as a case study – a once deprived inner city neighbourhood only a few minutes away from Berne's central station – we will approach some answers to these questions.

Working-class district Lorraine

Lorraine belongs to Berne's so-called inner ring around the city's historical centre. This part of the city was built during the late nineteenth and early twentieth centuries. During the past 25 years, it suffered a radical change: its working-class neighbourhoods became middle-class, with a marked increase in individualized lifestyles. More than half of the districts' households today are single households (Stienen 2007). During the past 30 years, in all big Swiss cities, a dramatic geographical shift of the urban working-class and immigrant population has taken place. In the late 1970s, working-class residents and foreign nationals were the majority in inner city areas. In 2000, however, these social groups were largely concentrated in the Fordist settlements built in the 1950s–1970s at the urban periphery (Statistikdienste der Stadt Bern 2006). The case of Lorraine highlights origins and local outcomes of such urban transformations.

Lorraine was planned and built by private investors as a working-class area in the late nineteenth century. Only the neighbourhood's southern part was equipped with an ensemble of higher and more representative buildings, containing larger housing units, offices and businesses; it lay close to the bridge which connects the neighbourhood to Berne's historic centre. The greatest part of Lorraine's urban formation was characterized by small brick and wooden houses of 3–4 floors which formed small, densely populated housing units giving the neighbourhood a village character. In the 1960s, the municipality acquired part of the neighbourhood's private-owned real estate and houses to demolish. At that time, the municipality planned to create an express highway going right through the district as part of the planned highways system which would cross Switzerland's capital to give an immediate response to growing mobility and expanding private traffic (Lüthi and Meier 1998). The city's planning authorities had proposed a total renewal of the area. According to the urban development paradigm of the 1960s, half of the neighbourhood's houses should have been replaced by modern apartment buildings. Areas for housing, work, traffic and leisure were designed to be strictly separated (Planungs- und Wirtschaftsdirektion 1970). But these plans were rejected by Berne's voting population and were never realized, so that the now state-owned real estate in Lorraine stayed untouched.

Up to the 1970s Lorraine was one of the most typical working-class neighbourhoods in Berne. It even got the image of being Berne's 'Little Italy' because a high number of Lorraine's residents were immigrants. Most were Italian and so-called guest workers who had been actively recruited in their country of origin to work in Berne's expanding labour market. Many were single since Swiss laws did not allow subsequent immigration of family members; they mostly lived in old, cheap and badly equipped flats or in small Mansards in the neighbourhood's state-owned houses, often shared with compatriots. The presence of immigrant guest workers who were doing low-paid jobs in industry and construction contributed to the upward mobility of the native working-class in Swiss cities (Hoffmann-Nowotny 1973). This process shaped Lorraine's socio-spatial development.

While the city's periphery changed decisively between 1950 and 1980, with the construction of new apartment buildings for the established native working-class, over decades Lorraine preserved its village character – during some 80 years no significant investments had been made in the neighbourhood. However, the municipality considered Lorraine a redevelopment area due to its physical and social structure. But by 1980, only a few new constructions had been realized in the neighbourhood (Stadtplanungsamt 1982). Up to that time, the old-boy network of the neighbourhood's traditional tradesmen and members of conservative and right-wing parties had been the leading voice of the neighbourhood. They had close relations with public administration and Berne's conservative government and provided their members unofficially and officially with information and favours concerning all neighbourhood affairs. The neighbourhood's old-boy network strongly supported the plans of Berne's conservative government to completely restructure Lorraine. The hesitant and only selective realization of the municipality's ambitious plans was not a matter of local resistance, but a consequence of the economic crises of the mid 1970s and the subsequent transformation of priorities in urban planning. In Lorraine, none of the planned projects were realized. Neither did substantial changes take place, nor was any attempt made to reverse the ongoing decay of the neighbourhood's private and state-owned real estate. The 1980s, however, brought wide-reaching socio-political and spatial changes to the neighbourhood.

Contested urban territories

In the 1980s, urban development in Switzerland was shaped by the rise of radical urban social movements, as was the case in cities all over the world (Touraine 1985). The countercultural urban revolt at the beginning of the 1980s in Switzerland has been called *Achtziger Bewegung* (1980s movement) or simply *Bewegig* (the movement). The Achtziger Bewegung rapidly gained ground in decaying working-class neighbourhoods in inner city areas where scrapped factories and decrepit buildings were squatted in order to build up Autonome Jugendzentren (AJZ – self-governed youth-culture centres), flat-sharing communities and self-controlled working places. The Achtziger Bewegung which also named itself ironically Bewegung der Unzufriedenen (movement of the unsatisfied), sought to shock self-satisfied urban Swiss society deeply, and to radically challenge the state with unlawful spontaneous cultural performances in public space, with squats and rallies which mostly turned into violent clashes with the police. Following the slogan, 'we want everything, right now' (Nigg 2001), the Achtziger Bewegung did not aspire to the 1968-generation's 'long march through institutions' to provoke cultural and social changes. Neither did it want to struggle for a remote revolutionary future to achieve what it was fighting for. It wanted to establish *FreiRäume*, 'liberated' self-controlled urban territories where radically alternative cultural activities and life forms could be developed 'here and now' in order to challenge the state as well as the city's establishment and its conception of what was 'normal' and socially acceptable.

Although the Achtziger Bewegung emerged in many Swiss cities at the same time and out of similar settings, it developed in different ways in each city due to local particularities. In Berne, the first AJZ was opened after squatting the 100-year-old building of Berne's former Reitschule (riding school) right next to the city's central railway station. In 1982, however, only one year after the squat, the city's police violently expulsed squatters, young people, moneyless artists, dropouts and other political activists of Berne's AJZ, and closed the centre by force.

Figure 21.1 Communal housing and urban regeneration in Lorraine, Berne. Once a typical state-owned house to be demolished, this dwelling was occupied in the
early 1980s, renovated step-by-step, and is today one of Lorraine's listed
buildings.

Photograph by Angela Stienen and Daniel Blumer.

The whole area was fenced in and given a 24-hour watch by the police over more
than a year. After these incidents, working-class Lorraine, which was only separated by a bridge from the closed AJZ, became a 'refuge' for many members of
the Achtziger Bewegung. They regrouped in the neighbourhoods' disinvested
and decaying real estate, after squatting some houses and renting others. Much of
the comparatively large amount of social housing owned by the municipality in
the neighbourhood was in such a bad state that it was relatively easy at that time
and place to apply for a cheap state-owned flat. At the time, Lorraine was considered the only place in Berne where the housing structure made it possible to
develop unconventional and alternative forms of urban life (see Figure 21.1).

In late 1982, some 400 people founded a housing cooperative and bought a five-
storey house in the very centre of Lorraine. In the house, a communally run restaurant, meeting rooms for political and cultural groups and political archives were
opened; two huge flats in the house were tenanted by two flat-sharing communities (Brasserie Lorraine 2001). All these transformations were closely watched by
the security police. The neighbourhood's old-established Swiss and foreign-born
working-class also observed suspiciously the changes in their neighbourhood.

During the 1980s, flat-sharing communities arose all over Lorraine and a wide-reaching communal infrastructure was built up, consisting of cooperatively run shops and restaurants, day-care centres, art and craft studios and political meeting points. The Achtziger Bewegung obviously became more and more territorially concentrated in Berne. At that time, it had a far-reaching impact in the city, where subcultural locations hardly existed, where restaurants closed at 11.30 PM, and where the 'other', (the unconventional, the unorthodox and the foreign) had been repressively banned from public space. By the mid 1980s, Berne's Bewegig had grown strong. The movement started to squat empty buildings spontaneously all over the city in order to organize one-night cultural events, which often attracted some thousand people, and it loudly called for a new AJZ in the city.

In 1987, after a period of urban unrest which followed the forced eviction of one of Berne's biggest squats by the police, ten thousand mostly young people marched on Berne's Reitschule, the place where the former AJZ was violently closed five years earlier. The empty and decaying building was again squatted and the AJZ reopened. From this moment on, hundreds of volunteers began renovating the building and turned it into a self-governed alternative cultural and political centre with concert halls, rooms for exhibitions and political events, a theatre, a cinema, a women's room, an info-shop for political groups (such as the squatters coordination), a restaurant and bars. In contrast to similar alternative cultural centres which appeared in Swiss cities at that time – such as the Usine in Geneva or the Rote Fabrik in Zürich (Wolff 1998) – the Reitschule did not receive millions in annual public subsidies. Most of the work that kept the Reitschule functioning from 1987 was done by unpaid volunteers. This situation has kept the centre's sustainability in permanent danger, but has also contributed to a strengthening of the centre's political legitimacy with the city's changing social movements. The Reitschule's operational community of interests – the Interessengemeinschaft Kulturraum Reitschule – is less exposed than others to blackmail by government financial cutbacks, and largely managed to defy governmental control of the centre's cultural and political activities. This may be a reason why the Reitschule, perhaps more than other cultural centres in Switzerland, is still today perceived as a centre for autonomous vanguard culture and as an important node of local, regional and national non-parliamentary political networks, even after 20 years of existence (Hansdampf 1998; Abteilung Zukunft 2007). Conservative and right-wing parties, however, have always considered the Reitschule Berne's 'blemish' and an obstacle to their plans for inner city renewal. Since 1987, the Reitschule is the city's most disputed inner area.

After the squat of the Reitschule in 1987 and the reopening of a self-governed alternative cultural centre in Berne's inner city, in Lorraine the situation grew tense. The political conflicts between the Bewegig and the neighbourhood's old-boy network intensified. The latter perceived the Reitschule as a centre of political conspiracy from where radical people were sent to Lorraine to infiltrate the neighbourhood and to deal out moral destruction (Blumer and Tschannen 2001, 2006). In fact, many political activists of the Bewegig who had settled in Lorraine at the beginning of the 1980s were involved in the squat. They largely contributed to its success, and were closely allied with activists of various political groups and networks of the radical left which had appeared all over the city and beyond Berne in the early 1980s. The Reitschule turned out to be the cross point where those groups and networks connected to each other and came to form the city's so-called *alternativszene* (alternative scene). After 1987, indeed, more and more activists of the alternativszene took up residence in Lorraine; they aspired to turn their everyday life into an integral part of political resistance. The territorial concentration of like-minded people in an inner city neighbourhood was considered a key condition to reach this aim. During the 1980s, the alternativszene became more and more a parallel society in Lorraine and provoked a shift of the neighbourhood's image from that of being Berne's 'Little Italy' to that of being the 'Bronx of Berne'.

Disputed gentrification

At the beginning of the 1990s, the situation in Berne changed due to a major political shift in the local government: for the first time in Berne's history, the conservative local government was deprived of power. In the 1992 elections, a red–green coalition (socialist and ecologist parties) received the majority of votes and has since then governed Switzerland's capital. The political shift to the left is not specific to Berne: during the 1990s, all the big cities in Switzerland moved to the left, politically as well as socio-culturally (Ziegler 2002; Hermann and Leuthold 2002). But Berne has always been one of Switzerland's most conservative cities and now is considered the country's most leftist city (Hermann and Leuthold 2004). In the 1992 elections, people from different political factions voted for change. Many opposed the conservative government's urban policies, which were considered retrograde, exclusionist and repressive. They called for a new urbanity, and policies that would transform the city and make it attractive for diverse social groups. In the elected red–green coalition, heterogeneous political interests converged.

Berne's alternativszene took advantage of the transformation of the city's political power relations and the political groups in the Reitschule radically claimed the right to affordable living space for all of Berne's low-income groups. They contested the city's severe housing shortage at the beginning of the 1990s, which was produced by increasing gentrification in many of Berne's inner areas, among other reasons. The revival of Berne's squatter movement challenged the city's newly elected government and helped provoke substantial changes in Berne's urban development policies. The new red–green government sought to fulfil the promises made during its election campaign and responded to the claims: first, it adopted the so-called 'Geneva-model' in the city: squatters would be evicted only if the owner of a squatted house could show an approved building permit and could prove that construction works would start immediately. Second, in Lorraine a substantial amount of publicly owned older real estate was given over to its inhabitants on a so-called building lease basis for 88–99 years. The possibility of building leases is a legal ploy in Switzerland's building laws. A loophole allows self-governed housing associations and cooperatives to 'buy' a house on a lease basis. Building leases mean that the state, as owner of land and – if this is the case – of the buildings built on this land, keeps the ground as its property but sells either the building on this ground or the building rights. The land is rented out for a fixed period of time, mostly for several decades (up to 99 years). After the rent expires, the state has the option to either buy back the buildings on its own land for the purchase value (not the market value), or to renew the lease.

In the case of Berne, another procedure has often been applied, which, in case of an early sale of the buildings by the leaseholder, gives the state the right to a first refusal on that transaction; the state's purchase option is treated with priority in order to limit speculation.

The self-governed housing associations that had been founded in Lorraine in the 1980s and early 1990s managed to tap the full potential of the legal ploy. Several building leases for state-owned houses have been negotiated with the municipality. The leased real estate has been used in different ways, and a variety of housing projects have been realized. They sought to respond to the growing heterogeneity of Lorraine's alternativszene. While some people continued working in the neighbourhood's self-controlled non-profit labour market or other low-paid jobs, others entered the formal labour market after finishing their university careers and often ascended socially. Partnership and family models diversified and so did lifestyles.

Although the housing projects responded to this diversity, all housing associations implemented strategies to prevent private profit making. They also sought to counter dominant socio-spatial trends in Berne such as individualization and atomization of the city's inhabitants and growing space consumption: urbanites were demanding more and more space for personal use as result of changing lifestyles (Stienen 2007; Statistikdienste der Stadt Bern 2006).

Lorraine's housing associations developed different strategies to reach these aims in the leased buildings. They imposed, for instance, a so-called Minus 1 Room Standard in order to limit the individual square metre use: in the buildings' flats only one bedroom can stay free and can be used as guestroom, all other bedrooms have to be occupied, i.e. as many persons have to live in a flat as there are bedrooms. Some housing associations used another strategy, pooling flats and promoting communitarian housing forms. Moreover, solidarity models were implemented to prevent people with low incomes having to move out of the leased buildings.

Lorraine's alternativszene did not resist regeneration in the neighbourhood, but tried to shape it. It managed to influence the decisions of the municipality on the development of wider construction areas in the district. In two crucial cases, it was able to ensure that the municipality prioritized investors' housing projects that accounted for the future tenants' preferences (see Figure 21.2) and constructed flats adapted to the particular conditions of the neighbourhood's residents (Blumer and Tschannen 2006).

The variety of housing projects which emerged in Lorraine during the 1990s and after the turn of the century reveals that the individual needs, preferences and economic possibilities of the neighbourhood's alternativszene diversified. The projects respond to a plurality of lifestyles, some of which are more alternative than others. For this reason, only few people of Lorraine's alternativszene who

Figure 21.2 Newly built housing in Lorraine, Berne. Housing project 5, a newly built housing development that accounted for the tenants' preferences. Part of the flats are inhabited by housing communities.

Photograph by Angela Stienen and Daniel Blumer.

initiated the transformations in the neighbourhood in the early 1980s left Lorraine and moved to other places in the city. Instead of being displaced, the 'pioneers' of Lorraine's gentrification established themselves in the neighbourhood and became firm and powerful contestants of its regeneration.

During the 1990s, Lorraine lost its image of being the 'Bronx of Berne'. Today, the municipality's urban planning and development office considers the neighbourhood a showcase for successful urban renaissance: 'a high quality of life, a diversified housing market which meets a variety of social needs' and 'a rich multicultural life' are the points which the office considers to be the district's hallmark (Blumer and Tschannen 2006:393).

The losers of Lorraine's 'success story'

The shift of Lorraine's image best symbolizes the far-reaching consequences of the described socio-spatial and political changes. The political change in Berne's government weakened Lorraine's old-boy networks. These networks lost their links with the local administration and thereby their influence both in the government and in the neighbourhood. In contrast, all of a sudden, it was Lorraine's established alternativszene who had allies in the city's government. Their influence in the neighbourhood increased and local power relations changed. The territorial entrenchment that was made possible by the negotiated building leases empowered Lorraine's alternativszene. It was no longer exposed to potential displacement. Instead of being the local government's adversary, as during the 1980s, the alternativszene turned into a respected (or perhaps co-opted) counterpart of the government. The confrontational politics of the 1980s were substituted by selective cooperation with the government, which regularly consulted Lorraine's self-governed housing associations on local development issues.

It might be argued that this story of contested and modified gentrification happened in Berne because Berne's alternativszene had been present both at the right place (in a neighbourhood where a large part of the decaying real estate had been state owned) and at the right time (the historical moment of transforming power relations in Berne's government) and also with the right legal concepts (to assume common responsibility for downgraded state-owned houses which the municipality could get rid of). Although Lorraine is situated very close to the city's historical centre, and had been considered an increasingly attractive place in the city, the speculation and regeneration that would displace the former residents had been impeded. Major reinvestments in the neighbourhood's real estate did not take place, and neither did cultural upgrading occur as a result of high-income gentrifiers moving in, as was the case in Berlin's Prenzlauerberg (Bernt and Holm 2002). Nonetheless, when passing over and through Lorraine, spot gentrification is very visible and the neighbourhood still is a disputed territory.

Who were the losers in Lorraine's 'success story'? The political changes in Berne had been determined strongly by the city's socio-cultural transformations during the past 25 years. The real losers of Lorraine's gentrification were the neighbourhood's once established working-class and lower middle-class to which the 'guest worker' population also belongs. These social classes perceived themselves as Berne's 'respectable middle class'. This is the typical self-image of Switzerland's working-class whose upward mobility during Fordism and its welfare system contributed to its socio-economic establishment. In Lorraine those residents considered themselves the guardians of the neighbourhood's norms of respectability. Whether native Swiss or guest workers, they perceived themselves as the neighbourhood's 'insiders' in Norbert Elias' (Elias and Scotson 1994) sense, as those who are conserving the real values of 'Swissness', against alien 'intruders' such as honesty, decency, law and order, private property, a sense of working hard, and unrestricted respect for state authority. These norms had been politically defended by the neighbourhood's old-boy networks. Those social classes had neither been displaced from the neighbourhood, nor were they impoverished. Nevertheless, they lost their political and socio-cultural hegemony in the neighbourhood as well as in the city as a whole, and were marginalized.

The people in this 'class' perceive the changes in their immediate environment as a threat to all that had been part of their 'lifeworld' and 'their' Switzerland and, in the case of foreign nationals, as menace to all that they once had aspired to when they decided to emigrate and work as guest workers in Switzerland. It is therefore hardly surprising that these social groups in Lorraine became followers of Switzerland's right-wing People's Party, the Schweizerische Volkspartei (SVP), which is the most powerful party in Switzerland's Federal Parliament and a member of the coalition government. They do not want to lose what they achieved during years of hard work and abnegation: economic stability as well as social and normative security, the issues the SVP promises to address. In Lorraine, however, these social groups were not openly contesting the neighbourhood's gentrification. They withdrew from the neighbourhood's public life and perceive themselves as politically powerless in a left-governed Berne (see Stienen 2006).

The struggle continues

In October 2007, a protest against supporters of the SVP ended in clashes and tear gas. A march of several thousand SVP supporters to the Parliament had been prevented by counter-demonstrators. In its national election campaign, the party used a poster which showed three white sheep standing on the Swiss flag as one of them kicks a single black sheep away. 'To create security in our land', the poster says. The message resonates loudly among Swiss voters, many of whom are convinced that Switzerland has become a haven for foreigners, including political refugees. Nearly a quarter of Switzerland's 7.6 million inhabitants are foreign nationals. The message of the party's political campaign is that the influx of foreigners has somehow polluted Swiss society, straining the social welfare system and threatening the very identity of the country: the SVP has developed a crude 'us-against-them' discourse. Although foreign nationals make up more than a quarter of the Swiss work force, Switzerland has perhaps the longest and most arduous process to become a citizen in all of Europe: candidates typically must wait 12 years and have to prove successful cultural integration and linguistic competence before being considered. After the riots which frustrated the People's Party's march to the Parliament, claims for 'cleaning' Berne's historic centre of the undesirable 'other' seem to be getting more of a hearing from Berne's red–green government.

There are great business expectations in the city with the hosting of the European soccer Cup Euro 08, in June 2008. Berne's most ambitious urban revitalization project at the very entrance of the city's historical centre – the upgrading of Berne's central railway station and the square at the station's forefront – has come to an end. It was inaugurated just before Euro 08 started. The project was expected to open a radiant vista of the city for the visitors and new investors expected to come in large numbers to Europe's 'most unimportant capital' in the wake of Euro 08. In this context, Berne's governing red–green coalition is assuming more of a neoliberal discourse of efficacy, austerity, security and order, aimed at attracting solvent taxpayers to live in the city. Thus, Berne's inner areas remain contested territories. Nothing better captures what this means than a graffito in Lorraine which reads, 'We don't perish of our defeats; we perish if we stop struggling'.

22 Searching for the 'sweet spot' in San Francisco

Peter Cohen and Fernando Martí

This is a story that revolves around San Francisco's South of Market (SoMa) neighbourhood as the development 'ground zero' for the past ten years. It is a bellwether of urban change in San Francisco both in physical evolution of the city landscape and in social implications of changing demographics, economic trends and politics.

San Francisco is a land-constrained city of merely 47 square miles at the tip of a peninsula marking the gateway to the San Francisco Bay Area region. It is entirely built out, with nowhere for new development to go except vertically or by 'recycling' existing land and land uses. While the pattern of gentrification is impacting many cities large and small, San Francisco is arguably at a more advanced state of urban gentrification than any other major North American city. The average new housing unit sells for nearly US$800,000, and the word 'luxury' has become so commonplace that marketers now use the terms 'super luxury' and 'beyond luxury'. The trend of 'pied-à-terres' (more benignly referred to as 'second homes') means that a substantial portion of the new urban class buying these condos are merely vacationing residents of the city. The city is *sold*, literally, as a playground of refined lifestyle and pampered living – a scan of the marketing ads for new housing projects in hot areas leave little doubt of this 'super-gentrification' (Lees 2003a) (see Figure 22.1).

At the same time, San Francisco has one of the largest homeless populations, and its working-class has been shrinking under pressure of job losses and expensive housing. Local-serving businesses and small industry have a hard time competing with the rise of chain stores and 'new economy' businesses. Aside from real estate development, city policy-makers are banking on R & D biotech and Internet businesses for economic growth, without knowing *who* will be put to work by those new jobs. Similarly, the mantra of 'the housing crisis' has resulted in policy-makers embracing any and all new development without discussing *who* the housing is for. The planning agencies, local media, and mainstream advocacy organizations concern themselves with issues of ideal urban form and 'liveability', represented by high rise 'slender towers' and streetscape improvements, while development continues to wreak havoc on the socio-economic landscape of the city. SoMa, more than any neighbourhood, has been a development wild west for the past several years (see Figure 22.2). A city Housing Inventory on development counts per area

Figure 22.1 Luxury condominiums in San Francisco.

Photographs by Peter Cohen.

Figure 22.2 Map showing SoMa development projects, San Francisco. The SoMa neighbourhood is contiguous with San Francisco's Central Business District, fuelling the intense pipeline of 'infill' development.

Map by Peter Cohen and Fernando Martí.

shows SoMa added nearly 4,000 housing units from 2001 through 2005, literally 'off the charts' by comparison to other city districts.

The end result is San Francisco squeezing and locking out the working classes, while a new urban upper-class infiltrates the city's old working-class and industrial neighbourhoods at an increasing pace.

In economic terms, this city's gentrification is tied to an excess of national and international capital sloshing across the globe and being parked in 'hot' cities like San Francisco. The cycle is further fuelled by the ever-enthusiastic interest in 'real estate' as a commodity for wealth-creation (property investment portfolios replacing tech stock portfolios in the post 'dot-com' bust), with condos purchased for speculation. The local government, starved for resources by an anti-tax State politics, is easily lured into becoming the facilitator of development capital as a source of local tax revenues and development fee tithes. In addition, the euphoric rhetoric of the 'smart growth' movement has helped fuel the market for urban in-fill development (this is an increasingly influential factor given the regional Bay Area's conservationist leanings). Environmentalists and transportation advocates have embraced a position of making accommodations for development as a strategy to achieve anti-sprawl conservation goals, and in the process have shifted environmentalism from a debate over growth versus no- or slow-growth to a pro-development 'smart growth' framework. In this growth-dependent economy of San Francisco, development policy tends to formalize what 'the market' wants to build, creating a vicious circle of real estate speculation and price inflation. As one local activist wryly puts it, land speculation is 'a civic tradition' in San Francisco.

The work of responding to the gentrification process in concrete ways is a hard road of incremental and patient chipping away at an entrenched paradigm of San Francisco real estate development. With *equitable development* as a core objective and 'public benefits' as a working concept for a redistributive mechanism, a variety of policy advocates have been making progress in recent years in shifting the policy framework to the left. *Whose city?* becomes the question of importance, not just generically how many housing units, or how many tax dollars, or how many new jobs. A combination of technical savvy and political strategy has incrementally advanced a set of more progressive standards for planning and development of the city that are struggling against the gentrification juggernaut. This chapter focuses on three key outcomes from these efforts: strong inclusionary housing laws, policies for public benefits zoning, and a strong city resolution guiding future planning and development.

This progressive policy work has come almost entirely out of the community rather than from the city bureaucratic apparatus. Once the politics are secured in support of an initiative, the officials are resolved to putting it into practice. There has been no grand plan, no designated resources, and no consistent strategy. Rather, this policy work is tied together loosely through a puzzle of intersecting processes and conversations and by the intentional cross-fertilizing efforts of some central players – a cumulative assemblage of advances.

As the experimental work continues to take shape in relation to immediate struggles, progressive policy advocates will face the ongoing challenge of trying to institutionalize this kind of 'public benefits' change within the system so that it sets a new bar to counterbalance San Francisco's gentrification.

SoMa: a brief history of gentrification and struggle

SoMa was the industrial backbone of San Francisco, beginning with the docks along the waterfront during the Gold Rush of 1849, then gradually building shipping-related industries, warehousing, and large-scale processing and manufacturing of everything from furniture to clothing to beer. Like other industrial areas of the United States, SoMa was sprinkled with enclaves of workers' housing in narrow alleyways, often near heavy manufacturing.

Due to its industrial nature and strongly unionized workforce, SoMa was the heart of San Francisco's history of radical organizing, from the late nineteenth century onwards. SoMa was the site of the most radical uprising in San Francisco history, the 1934 General Strike, led by the ILWU, the Longshoremen's Union. Throughout the twentieth century, SoMa has a very diverse residential community, due to its role as a place of arrival for immigrant workers and its connection to shipping throughout the Pacific. A Latino population developed in the 1920s and 1930s due to coffee and fruit shipping connections to Central America. A strong Filipino community took hold in SoMa to work in local food processing and manufacturing jobs. The African-American population grew rapidly during the 1940s with the dramatic growth of Second World War shipbuilding and related industries.

In the 1960s and 1970s, as heavy manufacturing jobs declined, residential hotels housed more older men rather than newcomers seeking jobs (the deindustrialization of San Francisco was already being planned as early as the 1940s, organized by major Bay Area corporations set on moving industry east across the Bay, and leaving San Francisco to function as a regional financial and cultural capital. The global restructuring of the 1970s completed the move of most large-scale manufacturing and warehousing out of San Francisco). Two large redevelopment projects in the 1970s and 1980s resulted in a large population loss in SoMa, with residential hotels becoming housing of last resort for people on the verge of homelessness. At the same time, the industrial grittiness of SoMa attracted a counter-cultural population, including artists, musicians and music clubs, a gay leather scene, and accompanying bathhouses.

Over the 1980s and 1990s some of SoMa's remaining industrial uses diminished; new big box retail, promoted by city policy desperate for tax revenue, moved in along its southern edge, near the freeways; a night club area and new large-scale redevelopment projects were built along the southern waterfront, culminating in a new ballpark, a new residential 'neighbourhood', and new ground floor retail uses comprised almost entirely of chain stores. While these development projects mostly affected large warehousing areas near the waterfront and along old railways, the speculation frenzy brought on by San Francisco's famous 'dot-com boom' of the late 1990s squeezed the area's many small industrial businesses. The new economy has resulted in a familiar bifurcation: the growth of a new class of high-wage skilled workers and asset-rich investors, and an ever-expanding sector of low-wage service workers, along with a loss of traditional blue-collar jobs.

Perhaps the most significant aspect of the anti-gentrification battles is the struggle for residents to remain in their neighbourhoods. This has primarily taken the form of a struggle to preserve the existing housing stock for the existing residents, to create significant protections for tenants, and to create additional housing stock that is affordable to them. A critical victory was the passage of a strong rent stabilization law in 1979. However, San Francisco's rent control allows that when a tenant leaves, landlords can raise rents to full market pricing (a provision called 'vacancy de-control'). During the 1990s dot-com boom, this gave motivation for creative ways to legally 'vacate' rental units, and rents would often triple in the process with what one might call 'new urban class' tenants. Rents have never returned to anything near pre-boom levels. That has made it clear that no single anti-gentrification tool can be expected to hold the line in the face of extreme market pressures as what San Francisco has been experiencing. Limited land availability is an important factor in the San Francisco 'setting', one that is often ignored by the local development boosters who argue that if the city could just 'allow' enough housing to be built, production would keep up with demand and therefore prices would stabilize and housing affordability would be a built in part of 'the market'.

The kind of market-rate development in SoMa can be divided into three models of condo development, marketed to different demographics. In the early 1990s, the gritty neighbourhood and large flexible loft spaces in the industrial areas attracted a younger 'creative class', working in dot-com start-ups, the entertainment industry, or as architects and designers, and students. In the late 1990s, dot-com businesses began to locate in industrial sites while even more affluent dot-com workers moved to the new live-work lofts and mid-rise apartments. Some of the newcomers worked locally, but many commuted south to Silicon Valley technology companies giving SoMa the character of a bedroom community and a car-dependent population.

Another trend has been the renovation of large historic industrial lofts and office buildings into luxury 'vacation homes'. Many of these displaced the garment shops that employed predominantly Chinese and other Asian/Pacific Islands immigrant workers.

The most recent stage is a new market for second homes and corporate housing, located in converted lofts and office buildings and new high-rise luxury towers. SoMa has seen it all – a complete evolution from a gritty back streets industrial past to the 'luxury living' hot-zone of the current development cycle.

Taking a different policy approach

'All politics in San Francisco lead to land use', that is, real estate development. This is one constant that has not changed much over the decades. The increasingly proactive and practical approach by progressive advocates to influencing city policy related to housing, development, and planning seems to be a change of course from past strategies. Now the question is: what *do* communities *want*, and how do we *get it* (compared with the previous: what *don't* communities want and how do we stop it...?).

A fundamental factor in this subtle shift is the diminishing role of government. Public programmes such as large-scale urban redevelopment and big planning initiatives are largely replaced by unfettered private sector development and market-driven public policy. The pressure on the communities comes from many fronts through a myriad of individual projects and bite-size masterplan areas – there are no longer easy targets or big handles for progressives. Redevelopment's chequered track record through the 1980s left a general mistrust of government redevelopment programmes and a reduced role of the agency in city policy. The late 1990s run-up in rents and evictions and the increasing 'property rights' rhetoric challenging government control have made clear the limitations of the city's rent stabilization and preservation approaches when the development market is 'hot'.

Government, even big-city 'liberal' San Francisco, almost by necessity is becoming more a facilitator of market-driven development policy than a driver of social policy. Government's roles as a redistributive intermediary between private wealth generation and the financing of public goods, primarily through tax policies, have weakened. Private capital is increasingly relied on directly for meeting public needs. The city negotiates with developers for public improvements, affordable housing and programme funding because public funds are simply not readily available. Progressive advocates are in more direct confrontation with the nebulous 'market' than ever before. There are obvious constraints to effective advocacy in this changing paradigm, but it also compels new creativity.

Another important factor is the return in 2000 of the legislative branch (the Board of Supervisors) to selection by district elections and with it a Board dominated by progressives of various stripes. Following the condo and dot-com development boom of the 1990s, many new Supervisors came into office on a slate of controlling development and countering the gentrification of the city's historic working-class neighbourhoods.

It is the sweat equity of a relatively small constellation of individuals that has been the engine for policy activism. Some newer and younger activists have joined ranks with seasoned progressive 'leaders' around land-use issues. As the configuration of advocates evolves, there are new perspectives and personalities influencing directions and strategies. The fact that San Francisco's physical and political landscape is really divided into neighbourhoods means that there is not always collaboration among advocates operating at a neighbourhood scale. These distinct communities can function like small territories based sometimes around race/ethnicity dynamics, historical relationships to city government or local organizing styles, resulting in uneven responses to similar issues around development and gentrification. Nonetheless, an informal 'Eastern Neighbourhoods working group' has evolved, functioning as a brainstorming venue and has created a laboratory for launching several specific policy initiatives, including ones outlined in this chapter.

Responses to SoMa's gentrification

In the panoply of activist responses to gentrification pressure, there have been many efforts, from public demonstrations to fights against development projects to ragings in the progressive press. Summarized here are three policy measures that have implications for SoMa and city-wide.

Inclusionary housing ordinance

San Francisco first adopted an inclusionary housing ordinance in 2002, requiring all market rate development projects to provide 12 per cent of their units permanently at below market prices without additional public subsidy. Although the policy was a significant measure towards balancing the housing market playing field, the pace of housing gentrification far outstripped any real mediating effect of the 2002 ordinance. In 2006, housing advocates worked with three progressive city legislators to amend the 2002 law to significantly strengthen its impact: the required percentage of below market units was increased to 15 per cent, or 20 per cent if the affordable units were built off-site, or if the developer chose to pay an in lieu fee. The ordinance requirements were also extended to development projects as small as five units in size and affordable units produced off-site must be within one mile of the market-rate development project.

These amendments aimed to increase the amount of affordable housing generated directly through the city's robust private development activity and increase the prospects for economic integration by ensuring low and moderate income housing is built in the same general area as new market-rate housing. 'Affordable housing' is defined across a very wide range of housing costs (in itself a policy debate). The 'affordable units' produced directly by private developers through the inclusionary housing requirements are aimed at higher points on the affordability spectrum than the substantial affordable housing units produced in the city by non-profit developers; however, in lieu fees are used to subsidize non-profit developers creating units at deeper levels of affordability. The geographic focus of the improvements is particularly important for SoMa where new condo development has rapidly outpaced the construction of affordable housing by non-profit developers. The details of the ordinance were hammered out through a technical advisory committee assembled by the city, comprising a politically delicate balance of representatives from 'the development community' (market-rate developers, attorneys and supporters), affordable housing developers and community development organizations.

It is important to note that the proposed amendments came from progressive advocates working the political process long before a formal advisory committee was set up to institutionalize their demands. San Francisco's inclusionary housing ordinance is now one of the strongest among big cities in the United States.

However, the city's true affordable housing need is for 64 per cent of all new units to be below market rate, based on State-mandated demographic projections of future job growth, at various affordability levels from the lowest-income rents for the neediest of the city's population to above-median income prices for working professionals. While a 15 per cent inclusionary requirement is impressive as a policy standard, a broader array of strategies is needed to get anywhere close to the 64 per cent equilibrium point that would hold gentrification to a manageable level. One piece of analytical ammunition has become instrumental: the city agreed to commission a study to answer a basic question: what is the impact of new market-rate housing development on the demand for below-market/affordable housing? The study's results demonstrated that every 100 market-rate condominiums units generate 24.94 lower-income households through the direct impact of consumption and a total of 43.21 households if total direct, indirect and induced impacts are counted in the analysis. Thus, new development in San Francisco creates a permanent lower-income workforce, for restaurants and cafes, child-care, gardening and home improvement, auto repair and so on. Read another way, private development has a responsibility, directly and indirectly, for mitigating the social costs of maintaining adequate housing for the workforce most vulnerable to gentrification impacts.

Public benefits zoning tied to new development

San Francisco officials are increasingly infatuated with the notion of 'densifying' the city as a 'smart growth' infill strategy for accommodating regional population growth and acting on public perceptions about a 'housing crisis', though, criti-cally, not clarified for the public as a lack of *affordable* housing. A 'new urban-ism' design approach to recycling older areas of the city's back-office and industrial districts is also in vogue. In practical terms, these concepts mean approv-ing generous increases in building densities, building heights and base zoning use allowances, and a desire to 'streamline' permitting of housing development.

Today's progressive advocates are generally not opposed to increased den-sity or scale of development, but focus instead on ensuring that the develop-ment is locally beneficial and affordable. They argue that density, height increases and regulatory streamlining in the name of smart growth simply con-fer value to property owners and developers without any clear 'public benefit' gained in exchange.

This analysis begat the idea of Public Benefits Incentive Zoning (PBIZ) – using the city's legal authority over setting development parameters as a mechanism to create real estate value that can be traded for various community-development needs. The greater the development allowances conferred, the greater the public benefits that should be extracted from the development capital. Developer inter-ests and the city administration responded to the concept as 'anti-development', and initial legislation in 2004 was shot down as too radical. But as advocates pressed the idea ahead and revised the technical details, it was increasingly seen by the city as a potentially rational way to harness San Francisco's lucrative development market. For progressives, it was an anti-gentrification strategy that could provide money for affordable housing construction and a host of commu-nity improvements in the face of rapid development that the city was essentially too broke to provide; for developers it could be a pathway to 'certainty' in the permitting process if their projects were not battled one by one and subjected to last-minute deals to win political approval.

The technically and politically tricky part of the formula has been figuring out how much of the value can be 'captured' while still allowing development projects to, as developers call it, 'pencil'. The development community's argument is that progressives could 'kill the goose that lays the golden egg', threatening that development may come to a grinding halt in San Francisco if public benefits tithes are too onerous. Many progressives would argue that if the egg is not golden enough, the community is left with just a goose. Thus, part quantitative analysis and part poker game, this debate over how far to push the edge of the envelope remains one of the key sticking points in terms of institutionalizing the PBIZ model.

The public benefits zoning concept is particularly relevant for the SoMa neighbourhood because of the significant 'upzonings' already made by the city, with more contemplated for the future. It is in many ways an ideal 'infill' area. It is also, however, an existing neighbourhood, with established working-class and people of colour communities. The intense development raises the question of what these SoMa communities are gaining from all this disruption and growth.

A test case for the public benefits zoning model was the struggle in 2006 to impose development impact fees on one sub-area of SoMa called Rincon Hill that was planned by the city as an enclave of high-rise condo towers. For many decades, the area had been primarily light industry, distribution warehouses, parking lots, and a few union halls (remnants of San Francisco's nearby working waterfront). SoMa community advocates began pressing for a development impact fee on the high rises. In response, a developer-friendly think-tank sponsored an economic study that concluded the Rincon Hill developments could afford only a US$4 per square foot fee. The city accepted the study with little critical analysis. Advocates challenged the assumptions in the financial analyses, including one that expected an excessive 30 per cent return on capital for equity investors. The local district councilman then engaged in negotiating an agreement between the community and the developers. The end result was agreement on a US$25 per square foot development impact fee. Thus, in early 2006, was created the 'SoMa Community Stabilization Fund' which will generate roughly US$22 million to address gentrification pressure in SoMa. The Fund will be allocated for a range of public benefits including affordable housing, youth resources, job training, homeownership assistance and small business development.

Since that test case, policy advocates have continued to refine the public benefits zoning model in expectation of hardwiring similar development impact fee requirements to all planning areas slated for new growth. They are also exploring creative financing mechanisms through tax increment capture to augment the funding that comes directly from the private developments, as most of the progressive advocates working on the public benefits zoning model concede that it is unrealistic to expect development fees alone to fund all the affordable housing and community development needs to maintain the economic diversity of neighbourhoods. The PBIZ response has built the argument that *both* private developers and city fiscal governance are accountable for mitigating the gentrifying impacts of new development.

Policy resolution guiding planning and development

Progressive advocates reinforced the notion of public benefits and equitable development when they got the Board of Supervisors to adopt a resolution in early 2007 articulating policy priorities around planning and development in SoMa and other neighbourhoods targeted for growth. The resolution was framed in populist terms, opening with, 'The residents and businesses of the neighbour-hoods within the Plan Arca represent an irreplaceable element of San Francisco's diversity, and their preservation as such is critical to retaining the very essence of San Francisco'. This is a reminder that the city's priorities are based on its people's needs which cannot be ignored to accommodate the desires of the development market.

Though it is non-binding on the city bureaucracy or private developers, this affirmative policy statement has become a potent reference point in debates at public hearings and in shaping the direction of development policy. At its core it is a reiteration of the overall affordable housing need for 64 per cent of all new units to be below-market rate to sustain the current and projected workforce of the city. It also goes beyond housing to address a range of services and amenities needed to retain that 'essence of San Francisco'. Among other things the resolution calls for 'identifying new transit routes and improvements sufficient to serve new residents and businesses while ensuring that the transportation needs of existing residents and businesses are met; … and … protecting light industrial ("blue collar") businesses, existing arts venues, work spaces and clusters'. It also reiterates the expectation that the city's development plans will provide 'a full array of public benefits to mitigate the impact of new development upon existing neighbourhoods', tying the resolution to the basic concept of public benefits zoning as a strategy to counterbalance gentrification.

The resolution passed unanimously through the 12-member Board of Supervisors, which includes a sometimes fragile super majority of eight progressive members and a few more conservative/centrist members who often side with development interests. Though 'the G word' was never employed, that resolution was a political tipping point that sent a clear message from the legislators to the city administration and department officials to reframe development policies in light of gentrification concerns.

All three of these initiatives outlined above are pieces of a puzzle moving the policy framework in San Francisco slowly towards a more equitable development outcome. Or at least that is what progressive policy advocates are aiming for.

Conclusion: is it making a difference?

It is too early to tell if these incremental progressive victories in the struggle against a super-gentrifying city are making a fundamental difference in the impacts of rapid development on the socio-economic landscape of San Francisco. The impact of SoMa's gentrification was evident in the challenging re-election race for the city's most progressive district Supervisor whose ward has seen tremendous condominium and luxury-tower development over the past four years. Despite negotiating the SoMa Stabilization Fund which will funnel million of dollars into his district, he nearly lost his seat due to the shift in voter demographics. Only a massive organizing campaign among city-wide progressives kept him in office. As a local saying goes, 'who lives here votes here' – the ultimate stake in the gentrification struggle is the political configuration of the city's body politic.

Will these policy responses maintain the diversity of communities in SoMa? Can we turn the screws even harder? The intensified level of action raises the question of whether increasingly vigorous progressive policy work is sustainable over time given the limited advocacy 'infrastructure'. Is there a stable of human and technical resources for ongoing analytical and logistical capacity to keep raising the bar? The 'other side' in the struggle, the incessant real estate development that drives gentrification cycles, only oscillates but never loses momentum. Are we trapping ourselves by thinking that we can 'manage' gentrification?

A handful of progressives are questioning the entire notion of trying to find the 'sweet spot' between the interests of private sector gentrification-inducing development and preservation of an equitable, diverse, politically progressive city. 'Public benefits' depends on supporting capital, bringing in investment (which these days means extremely fluid global finance) to pay directly and indirectly for public needs. Can a strategy of mutual accommodation ever match the growth machine or create 'equitable' development policies? These advocates are inclined towards a system of firmer limits on growth coupled with public benefits policies – treating the city and its people as the assets, with development given the *privilege* of investing in its growth and improvement on terms more explicitly spelled out by and for the city's communities. That is a radical notion considering the culture of real estate development as the engine driving public policy. But if the screws are to be turned as hard as they can in the broader struggle to 'do something' about urban gentrification, there is perhaps no better place to try than in San Francisco.

Bottom up vs top down

Jess from Pepys, a social housing estate in South London

Mark Saunders

Jessica Leech has been a Pepys Estate resident for 16 years and is a housing and community activist. She is chair of the Pepys Community Forum, a Single Regeneration Budget (SRB) funded community-led organization.

Jessica Leech.

Photograph by Mark Saunders.

'Round about 1997–98 a group of residents and local community workers became really concerned because the area regeneration funding was running out. It was really clear that the council weren't going to continue to provide services and while the estate here had undergone housing investment there had been no investment in any of the social and welfare infrastructure.

So a group of about 15 of us decided to put in a bid to the Single Regeneration Budget, and our bid was entirely about developing the capacity of various community facilities and that the community would be at the heart of deciding what was important in terms of [improving] local people's quality of life and opportunities.

And we also wanted to move from a situation where things that were working well for the community ceasing to be funded because it was no longer flavour of the month.

So we put together this SRB bid for 3.5 million pounds, and we were very surprised because the Government Office of London were very interested.

At the same time Lewisham Council also put in a bid for the physical regeneration of Silwood. The Silwood estate had been neglected for a couple of decades and was really in need of some investment. And so the government were very interested in spending some of the SRB money on refurbishment of the Silwood but felt that the council's bid, because it didn't have significant community element at that stage, was too weak to succeed.

The Government Office of London also had concerns about us as a group of small community organizations and local residents actually being able to handle … a large government bid in terms of accountability, handling the money and things like that so the government came back to both Lewisham and us and said neither of you will be successful in your bids, unless you merge and put in one bid.

Both sides felt they had to agree, but it was never an easy combination because I think the philosophy that laid behind both was so contrary, because we were very much about putting residents in charge of how the money gets spent; residents were in charge of the budget, residents were the people who were making choices about what services were brought in. Whereas on the Silwood most of the investment was around the housing and then it was the local authority saying these are the targets we've got to meet, who can we bring in to satisfy these objectives.

We had hours of debate about whether or not we wanted to join forces with Lewisham because of what that might mean in terms of losing some control.

And so even though we joined with reservation we knew what we were getting into and the community had said, "yes we're up for this".

And as time went on [Silwood residents] felt cheated, because it was never explained to them about the fact that our [Pepys] money was ring-fenced, because again they didn't really understand that we were part of the package, right from the very outset. And so I think that led to a lot of tensions that needn't have been there.

The first few years of our relationship with the Silwood SRB was very tense, extremely tense at times. Lewisham was really uncomfortable with having signed off 3.5 million pounds over seven years to this community organization, who were just going to do what they wanted.

They just really wanted to control the process, and we weren't prepared to allow them.

Eventually, almost like a truce was called. Silwood SRB just decided that, apart from dealing with our paperwork and making sure that the processes that we were using had integrity, just to simply ignore us. And so from about year four onwards it was almost like we were written out. We had our money, we did what we did, they did what they did. I communicated with the project officer and the finance officer, but that was it.

For us that was a huge relief, but I also think at that point, something very important was lost, that we would pilot a lot of the ideas, that we would be a bit of a social experiment about what happens when you put a community in charge of something, how does a community actually achieve sustainability for the services.

I think that the Silwood residents and the Silwood bid lost a huge amount of potential learning and knowledge.

One of the things that we were able to do that I don't think the Silwood were able to do is if a resident came to us about something that needed to be done, or identified a gap in something … then even if it took a couple of years for us to make something happen, we were able to take that concern and make it into something. So there are things that we are still doing now, that came from local residents.

A refugee employment advice project, our composting project where we collect food waste from all the households, … the Splash Water Sports centre where young people are learning how to sail and canoe, … all those things have survived the SRB, all those things are ideas that have come from residents and what we've been able to do with it has just never been replicated on the Silwood.

We invested a lot in Riverside Youth Club, a project called CACOA, they produced a very high quality (youth) magazine (*comfusion*) for young people … a lot of those young people were involved in that project and gone on to do very well. We also spent money on PEPYS resource centre, Joblink, and a CV service.

We funded Deptford and New Cross Credit Union which is about cheap saving for people on low incomes. And we developed a food co-op because there was no fruit and veg place on the estate. There was also developing the community garden, on a piece of unused allotment land. And the idea of people beginning to grow food (fruit and berries) that they can take home with them but also as a bit of an educational tool. We developed a recycling and composting project where we collect food waste from the flats, the local authority won't do door to door collection from flats in Lewisham, but we also did quite a lot of interesting smaller stuff.

There is too much to remember really.

[For successful bottom-up regeneration] I think you need a space where people feel comfortable going … so that must be a space where not every minute is booked out by some service provider, but a space where people can just drop in . . ..

You need to resource staff so it's not constantly about, "you've got to deliver these outputs", but that actually, I've got time to sit and chat, because people need to be listened to. People aren't always very good, in a few minutes, saying exactly what they mean to say. I think you need to be able to support people building relationships with other people in their community as well.

Because when people begin to connect, they then begin to be able to talk about their environment. And then if there is somewhere that people feel they can take that conversation to, people then don't feel as if they're "I'm the only one that thinks this way" or "that's a bit scary because I've got no one else backing me up".

You need community events, so people have the opportunity to see who their next door neighbours are, because they then begin smiling, then it's "good morning" and it builds, and, I think, starts to provide people with a stake and a concern about where they are and what's going on around them.

But those organizations that do hold the purse strings, or do make the decisions, also need to be in a place where they are willing, or able, to allow the community to genuinely affect the decision making. So when people talk about community development they talk about bottom up. They're always talking about what the community needs to do, but actually I think it's not just about what the community needs to do. It's about what other people need to do as well, in order to make that happen, in terms of changing their thinking. And what was just so unusual about the period of the SRB was just this strange thing where the community was also the organization that had the power to make the decisions about where what was a fairly significant sum of money went.

So we always had very open bidding rounds and we were very transparent about how and why we were making the decisions, and we developed, with members of the local community, strategies. Once we had the strategy, we then used that as the basis for making decisions about what we going to fund.

So we were completely open about everything; it didn't stop some people from feeling bad about us but it did mean that people could see how, and why we were making the decisions that we were making and I think that we did a lot better than say sometimes the local authority does.

We had an experience two years ago whereby the council invited applications for summer schemes and didn't tell people what the criteria was for making the decisions and also seemed to move the goalposts.

The early days of neighbourhood renewal funding, as well, that seemed to be very stitched. And in fact both processes only changed after considerable campaigning and outrage, largely from the voluntary sector. So in terms of excluding people, or making bad decisions or having poor processes, I don't think the local authorities necessarily get it right every time.

What tends to happen is the local authority will write the strategy, and then ask the local community what it thinks, whereas what we were doing was asking the community what it thinks and then writing our strategy, but it actually takes quite a lot of financial investment.

Most often the most vociferous and active within a community are those people who don't have families. I can think of the four most active people on our estate, are all single, are all sixty, or almost sixty and over, one of whom works part-time, the other three don't. But who've worked all their lives. So instead of working they just do community activism and all that they do is very good but because they've all only ever been single, because they're all white, and that relates to their age and housing patterns on the estate, because they're of a certain age, what's important to them isn't necessarily what's important to the community as a whole, and yet they're the people with the time to give the voice.

Unless you have somebody, you're actually paying someone within an organization, to knock on doors, to talk to people, it is very easy for you to become representative of only a small group.

There were a couple of times when we've done door-to-door surveys and we've paid local residents to go round and talk to other local residents. Instead of bringing in a company [of consultants] you get the survey information but you also get relationship building because you've got neighbours talking to neighbours.

The activism on the Pepys had a different dynamic to the activism on the Silwood.

At that particular moment in time, because the money arrived just at the right time, the Pepys were in a good position to take advantage of it. But the reality is that I have met on the Silwood a whole range of people and that same dynamic could be recreated, the personalities are there, the people are there. It's about what is there that is pulling it together.

From what I see and pick up is there is nothing taking that somewhere, and bringing that together, it's a mechanical approach [there]. Not about creating a change. And I think that probably is because London and Quadrant is not interested in losing control, for whatever reasons.

For me as a local resident, being part of the group that initiated the bid and then somebody who became employed by PCF its been very hard work, but also I still feel, nine years on, that everything I have done has been hugely worthwhile and has actually made a difference. And I am just as committed to it now as I was when we started.'

Part V

New theoretical and practical insights for urban policy

23 Whose urban renaissance?

Libby Porter

Each story in this collection is unique. Yet there are distinct patterns that suggest the specificities are actually variations on a theme: urban regeneration as a spatial economic restructuring of city neighbourhoods through reinvesting in disinvested spaces. This reinvestment has varied consequences in different places and on different people, contingent upon historical circumstance, geography, political and institutional cultures and socio-cultural values. Yet the consequence of rising land values – a ubiquitous intent of urban regeneration policies – is near universal. Those most vulnerable to the restructuring of urban property markets are hit first and hardest.

In this chapter, we put the stories in this volume into conversation with each other to produce some more general insights into the phenomenon of urban regeneration. By returning to the questions posed in the Introduction, we look at what kinds of regeneration policies are at work, what they look like, how they function, and their consequences. Our findings take issue with the literature that heralds uncomplicated benefits from urban regeneration (Bianchini and Parkinson 1993; Landry 1996; Roberts and Sykes 2000; Carmona 2001; Ginsburg 1999). We find there are benefits, but these rarely flow to those who stand to lose most from urban regeneration.

Let us be clear about our approach here: the journey we take is not one of raw comparison. Problems abound with such a project (see Carpenter and Lees 1995; Smith 1996; Slater 2004b; Shaw 2005b), and we have no claim to a universal theory or a watertight mechanism for comparison across vastly different contexts. Authors contributing to the collection have taken their own approach to reporting the stories arising from their own research. While we provided some parameters to authors for their contribution, we set no unifying methodology. We cannot, for example, compare data on displacement between cities (see Atkinson 2002, Shaw 2005a and Lees *et al.* 2008 on the problems of measuring displacement). Instead, our approach has been to allow authors to tell the 'stories of personal housing dislocation and loss, distended social networks, "improved" local services out of sync with local needs' (Atkinson and Bridge 2005:2) they have each uncovered. Different interpretations are always possible.

While we have deliberately avoided imposing a very rigid comparative structure, nonetheless some kind of framework is required to generate conceptual clarity. This has been an organic process, the themes evolving from the stories themselves. The categorization is entirely ours and could of course be differently constructed. It offers a starting point for further critical debate about urban renaissance policies and allows us to synthesize the consistent and robust themes that emerge to create insights for further debate, research, and policy action.

Our framework begins by looking at what kinds of state action are involved in each story. As Wyly and Hammel (2005) put it,

> More than ever before, gentrification is incorporated into public policy – used either as a justification to obey market forces and private sector entrepreneurialism, or as a tool to direct market processes in the hopes of restructuring urban landscapes in a slightly more benevolent fashion.
>
> (p. 36)

Building on their approach, we conceptualize two broad, fuzzy-edged forms of state action. First are *market-directing* strategies where overt, noisy forms of state intervention respond to market 'dysfunctionality'. Second are *market-obeying* strategies, where state interventions have a more discreet, even silent role, albeit for precisely the same ends. Of course, strategies that seek to direct markets are always already obeying and supporting them, so the categories are not mutually exclusive. Similarly, we acknowledge that states are neither homogenous entities, nor unified in their intent. States operate at varying spatial scales, according to various interests, and in a variety of institutional constellations.

Underpinning the variety of policies, strategic interventions, and modes of regulation discussed in this collection is a broadly conceived neoliberal ideology that 'privileges the unitary logic of the market' (Peck and Tickell 2002:387). We find the policy mechanisms described in the book to be geared towards an end-state of wealth accumulation, distributed and organized through free market logic. As noted in the Introduction, we see the urban renaissance policy dogma as deeply implicated in the construction of particular urban places as 'deprived', and that such deprivation is an improper part of urban life (Baeten 2004). Urban renaissance policy interventions, then, are designed to restructure and revalorize urban space, and usually target those places with the most marginalized populations. Yet, the 'actually existing neoliberalisms' (Brenner and Theodore 2002b) presented in this volume are diverse and we find the differences and specificities really matter. Our attention, then, is squarely on the multiple 'techniques of neoliberalism, the apparently mundane practices through which neoliberal spaces, states, and subjects are being constituted in particular forms' (Larner 2003:511).

This chapter is structured around these two categories, and sub-categories within them, and pivots on a set of analytical questions that we applied to the stories in each category: what are the narratives of intent underpinning the strategy, how is the strategy constructed and implemented, what are the outcomes, and who stands to benefit? Our analytical attention is focused on the specific mechanisms by which strategies are imagined and implemented. We are *most* curious about what the contingent parameters produce: particular urban forms, the existence of struggle, cultures of policy-making, and shifts in local housing markets.

Market-directing strategies

Cities are heralded as the engines of economic growth (Parkinson *et al.* 2004; Turok 2005), and the entrepreneurial turn in urban governance (Harvey 1989b) has pitched policy efforts towards creating the conditions for new competitive niches to flourish. As the distribution and nature of land use in the inner city is fundamental to this growth agenda, local states arrange their governance mechanisms and policy approaches to coordinate urban development for competitive ends. This collection rehearses a well-known set of economic sectors where city governments increasingly look for economic salvation: financial, business and knowledge economy sectors, creative industries, and tourism all feature here.

Istanbul, Johannesburg and Riyadh best exemplify just how virulent the economic competitive turn in regeneration policy has become. After looking at how these work, we look at specific sectors which lead particular urban regeneration policy approaches (tourism, culture, housing) and then close the chapter by considering the potential of grass-roots action.

Cities as the engines of economic growth

A particularly revanchist (Smith 1996) brand of urban policy is currently being deployed 'off the shelf' of western cities in places such as Istanbul, Johannesburg and Riyadh. Each is being shaped by city governments pursuing an agenda of rapid modernization to create their cities as 'world class'. The policy intents are clear. To be a competitive world (or even regional) economic player, a city must have space in its core to accommodate business and financial services, as well as a modern, wealthy image.

Inner city poverty is a major barrier to this goal. Regeneration is required to clear the inner city of the poor and allow capital a freer reign to provide the office, service and residential space to attract footloose global capital and the 'right kind of resident' as the City of Johannesburg policy states (quoted in Winkler, this volume). In each of these cities, the state is actively involved, indeed coordinating, the assembly of land for the private development market.

However, the techniques differ. Public–private partnership has been the dominant coordinating arrangement in Istanbul. The central government Mass Housing Administration (MHA) has the power to partner existing companies to 'regenerate' squatter districts, and 'revenue-share' where income from profit-oriented projects undertaken in partnership funds other activities. The City of Johannesburg is pursuing a form of 'aggressive reregulation, disciplining, and containment' (Peck and Tickell 2002:389), through tax subsidies in Urban Development Zones, and the wholesale eviction of residents. In Riyadh, however, the establishment of the Ar-Riyadh Development Authority as a 'development arm of local public authorities' (Ledraa and Abu-Anzeh, this volume) has failed to stimulate the kind of private investment and development hoped for. As a result, the public purse in Riyadh has underwritten all of the projects undertaken in the inner city as it waits for private investment to get interested.

Implementing these diverse policies has had dramatic consequences, especially if you are one of the urban poor standing in the way of the city competitiveness dream. Thousands, possibly tens of thousands, of people have either been displaced already, or are in danger of being displaced, across the three cities. Winkler estimates as many as 25,000 tenants may be displaced through the Better Buildings Programme in Johannesburg. Gündoğdu and Gough project that more than 2.5 million (mostly poor) people may be forced to the periphery as the MHA continues its plans to demolish around 60 per cent of the settled area of Istanbul. Estimates of displacement are not available in Riyadh. However, Ledraa and Abu-Anzeh point to the vulnerability of those only holding tenancy rights in Riyadh's urban regeneration zone.

The outcomes of this particularly aggressive form of property-led regeneration strategy are clear. Inner cities are radically restructured – economically, socially, demographically – by state intervention. The intent is clear enough (move out the poor to make way for capital to revalorize space), but the methods are diverse and contingent upon local conditions.

Tourism-led regeneration

Successful tourism-led regeneration policy hinges on the extent to which city governments and local urban elites can sell an authentic city identity, based on physical monuments, cultural heritage and place qualities. If only a city can establish its 'brand', the tourist industry offers a way out of poor economic performance. The intention is to maintain (or if you do not already have it, establish) a distinctive heritage offer in the inner city, by physically revitalizing city spaces.

In Florence and Rome, where tourism has long been integral to economic growth, city governments are orienting their policies towards the interests of short-term visitor populations. This often requires the removal of existing uses and residents from the city centre, and reorienting land use across the metropolitan area. Where an established tourist 'offer' does not yet exist, such as in the Ruhr Valley and Salvador da Bahia, the local state must create it.

How does this policy form manifest in these diverse contexts? Mega-projects are considered in the literature as fairly typical in tourism-led regeneration (Balibrea 2004; Vicario and Monje 2005). Our stories here, however, suggest that it is not only mega-projects, but also the smaller day-to-day planning management of inner city places, that can shape their transformation. Policy attention in Rome and Florence is on the small permissive ways space for tourism is created. One by one, residences are turned into small hotels, open air markets close to make way for boutique shops, and piazzas are overrun with cafe tables. The local authorities in both these cities express a melding of 'quality of life' interests with the needs of the all-important tourism industry as central to their planning frameworks. Yet these potentially more progressive and participatory elements are often overshadowed by the imperative of tourism growth.

In Salvador da Bahia, the local state is more actively and obviously integral to the transformation. The Monumenta programme sets out to restore the historic assets of the inner city, building by building. State-funded cultural animation projects are employed to attract visitors. For the architects of the Ruhr Valley's IBA Emscher programme, mega-projects are part of the reinvention of a massively deindustrialized regional landscape. Yet these are coupled with smaller, seemingly more organic, interventions that emerged through the deliberations of the IBA Emscher Park Planning Company.

The outcomes and consequences of these various tourist-led regeneration strategies are diverse. Rome and Florence continue to be successful in tourism terms, though at the cost of considerable displacement from the inner city of existing residents and uses. Soaring land prices in Rome and the orientation of metropolitan-wide planning policy on land use and transportation in Florence underpin this displacement. Significant displacement has also occurred in Salvador da Bahia. Tarsi (this volume) reports a vigorous programme of evictions where landowners are offered stark choices and more vulnerable tenants evicted. Making space for increasing numbers of tourists to be entertained, dined and accommodated requires the displacement of existing residents. Such 'tourism gentrification' (Gotham 2005) clearly shows how the global tourism industry intertwines with the production of locally specific symbols to become a distinct form of gentrification produced by urban policy.

Displacement, however, has not been a feature in the Ruhr Valley which has seen some success, in tourism terms, for the 'industrial monuments' projects. But in the Ruhr Valley case, there was nothing to displace, or rather it had already been displaced long before as industrial capital moved its operations elsewhere. Müller and Carr lament the lack of real economic success of business start-ups in the IBA Emscher Park. But we wonder, from a different cultural vantage point than that of Müller and Carr, whether this moderates increasing land prices and assists in avoiding the negative consequences of regeneration strategies. There may be much to learn from the Ruhr Valley case to the extent that a more modest approach, with incremental decision-making structures and space for non-economic projects to flourish, seems not to produce the aggressive class restructuring of gentrification. But places like the Ruhr Valley, which will surely feel the pressures on local land markets eventually, might also learn from other places where those pressures have been effectively managed. We return to this idea in the final chapter.

The outcomes of tourism-led regeneration, then, are highly locally contingent. The existing tourist 'offer' of a city can be crucial as it may determine the extent to which urban governance is already directed towards continually recreating the city as a tourist destination. Tourism-led strategies seem rarely to favour the interests of local people, as a highly sanitized version of a city's identity is preferred over the messier reality of the cultural lifeways of citizens. Such disjuncture can prove productive when the local citizenry actively struggles against the 'poster-child' mentality of tourism-led regeneration.

Culture-led regeneration

Even before Richard Florida (2002) had invented his creative city index, harnessing the symbolic capital of culture in cities had become a central feature of city development (see Zukin 1995; Harvey 1989c; Scott 1997). Linking culture with place, mostly through regeneration-type strategies, has spawned a whole set of activities called 'cultural planning' (see Bianchini and Parkinson 1993; Landry and Bianchini 1995; Montgomery 1995; Evans 2001), and a concomitant critical academic literature (Miles and Paddison 2005; Miles M. 2005; Evans 2005; Miles S. 2005; Shaw 2005b; Porter and Barber 2007).

In this category, we call together the stories of Amsterdam, Barcelona, Birmingham and Berlin. Artists and other cultural producers are present elsewhere in this collection: Melbourne, Toronto, London and Berne in particular. In this section, however, we focus on *culture-led regeneration* as a state activity, rather than the presence of artists as pioneers of regeneration (and gentrification).

Economic development and urban policy-makers in Berlin, Birmingham, Amsterdam and Barcelona perceive creativity and culture, however defined and imagined, as an engine for growth and wealth creation. The logic in Berlin and Amsterdam is to build on an existing 'brand' of creative city to attract further investment. The Media Spree Regional Management company in Berlin used (often cynically, according to Bader and Bialluch) the existing cultural cachet of the district to bring about major waterfront projects, none of which do terribly much to support the existing subcultural groups, without whom a 'cultural quarter' would not even exist. In Amsterdam, the intent at least is more benign, where the *broedplaatsenbeleid* (BPP – breeding place policy) actively supports subcultural groups by subsidizing 'freespace'.

The planning approach in both Birmingham and Barcelona's Poblenou district, however, is to largely ignore existing creative industries and attract new ones instead. Martí-Costa and Bonet-Martí illustrate how the '22@bcn' plan sought to displace existing users, demolish buildings and make way for a new 'knowledge society' production district. Similar mechanisms are at work in Birmingham, where the local state, in conjunction with its Regional Development Agency partner, is assembling land through compulsory purchase for a new high-tech, creative-led, knowledge economy 'quarter' of the city (see also Porter and Barber 2006, 2007).

Creative elements in cities can be both the pioneers of regeneration, tipping to gentrification, at the same time as being a critical force in urban struggles. In both Amsterdam and Barcelona (not to mention Melbourne, Toronto and London), artists have been at the forefront of direct anti-gentrification action. Our stories serve to illustrate a well-established literature on this point (Ley 1994; Ley and Mills 1993). Yet the outcomes are far from certain. Where cultural producers are involved, the tendency to displacement seems less vigorous. In Berlin, this is because there is enough existing space in the district to accommodate those moved on by the major projects. Gentrification has barely begun here, though it is probably only a matter of time. Locally grown cultural producers are at risk from the influx of the global creative class. Without policy intervention of a different kind, the cycle will turn, investment will boost, those alternatives will quietly disappear and gentrification will result.

Plan-led approaches, where a spatially oriented 'fix' of creativity is proposed for a particular urban neighbourhood, such as in Barcelona and Birmingham, complement a business-as-usual approach. Creativity or culture is nothing more than a particular sector of economic activity. Somewhat perversely, those city governments seeking to 'invent' creative quarters often do so by actively displacing artists and subcultural groups from those very neighbourhoods. Where greater attention has been paid, such as in Amsterdam, to the existence of subcultural groups and the paradoxical nature of planning intervention in the kinds of spaces they occupy, there seems to be greater potential. Tensions and dilemmas will always exist. The question is whether urban policy-makers are prepared to attend to those paradoxes in productive ways.

Correcting housing market failure

A specific manifestation of market-directing strategies occurs in those city-regions with 'dysfunctional' housing markets. This dysfunctionality is varied: in England's South East, the parameters are overheated demand, limited land supply and soaring prices; in Leipzig, the problem is oversupply of housing with lack of demand; in Mexico City, depopulation is causing major environmental problems on the periphery as well as entrenching urban poverty in the core. Yet in all these cities, the strategic interventions of governments are embedded in a socially oriented rhetoric of regeneration, where 'balance' must be restored to the market (or in the case of Leipzig, a market must first be created). In none of these cases is the state itself building social or other affordable housing units.

How do these housing management strategies work? The Sustainable Communities Plan of the UK government is a housing growth strategy that stimulates house-building activity by building essential new infrastructure. In Leipzig, the market is stimulated by reducing supply, through demolitions, and encouraging demand through a kind of 'training' of the population in the value of owner-occupation (Bernt, this volume). Mexico City's metropolitan government takes a far more regulatory approach to specifically control where the market invests and builds (García-Peralta and Lombard, this volume).

The intent of raising land values, while clearly present in Leipzig, is less present in the UK case (in a context of a massive affordability crisis) and in Mexico City. Interestingly, the consequences do not match the intent. Bernt (this volume) shows how ineffective the Leipzig policies have been in creating a broad general upturn beyond some areas that were already beginning to gentrify. Colenutt (this volume) argues that far from solving the affordability crisis, the UK Sustainable Communities Plan will simply polarize existing and new populations and cause exclusionary displacement. In Mexico City, the consequences have been immediate: soaring land values in the boroughs supported by the Bando Dos by-law, the displacement through price of existing residents, and an even greater pressure on the metropolitan periphery than existed before (García-Peralta and Lombard, this volume).

Land value increases have resulted where no specific policy intent existed, and have not resulted where they were intended. Why? Bernt shows that in Leipzig the major limits to the effective realization of the policy were landlord non-cooperation, a flagging economy, and a series of tax law changes that (regressively) concentrated urban property in the hands of larger property developers. While Bernt points to the general ineffectiveness of policy, we wonder whether in a context of better market conditions, this policy would have been more effective and potentially far more dangerous in its gentrification tendencies. As in Berlin and the Ruhr Valley, we can identify hope for containing gentrification where there is modest, and not rapid, economic growth, and continued fragmentation of land ownership.

The Bando Dos by-law in Mexico City had perverse policy outcomes. As it attempted to halt de-urbanization, it actually increased that trend, but with more socially regressive implications as the poor were pushed further to the fringe. As it attempted to provide more land for housing and make prices more affordable, it in fact concentrated land ownership, drove up prices, and reduced affordability. Rather obviously, we suggest policies need to take account of the actually functioning and real market processes within which they will be implemented (Gordon 2004:375).

Policy design elsewhere (Birmingham, Riyadh, Istanbul, Salvador, Barcelona) shows concerted efforts to concentrate property ownership and development rights in regeneration areas. This policy feature should be a target for future urban struggles, and alternatives to 'comprehensive redevelopment' through land assembly would be a fruitful focus for further research and policy debate.

Urban deprivation and social exclusion

In among the economic competitiveness mantra of urban policy, a strand of policy focused on area-based regeneration initiatives with a more socially oriented 'cohesion' and deprivation agenda still persists (see Atkinson 2000). The story here is a familiar one: districts and neighbourhoods within cities become marked as 'deprived' because they are populated by groups that consistently under-rank the average population on a range of social indicators. Physically, these districts are considered 'deprived' due to poor and overcrowded housing conditions, isolation from services, transport and employment centres, and damage to the built environment through persistent vandalism.

Deprivation-focused regeneration initiatives are particularly prevalent in the United Kingdom and the United States, where the 'discourse of community' in urban revitalization has had considerable purchase (see Imrie and Raco 2003 for the UK context; and Imbroscio 2008 for the US). Housing Market Renewal Pathfinder Initiatives in the United Kingdom and the HOPE VI programme in the United States fix on a range of social pathologies and their (presumed) physical manifestation through demolition and rebuilding of social housing estates. The destructive nature of these programmes, their socially and geographically constituted knowledge bases and their persistent inability to deliver benefits to the poor are now being exposed (see Cameron 2006; Allen and Crookes 2008; Imbroscio 2008; Allen 2008).

In whose interests is this form of regeneration enacted? The policy rhetoric is always geared towards a positive effect on the poor, or in contemporary language the 'socially excluded', though as Cameron (2006:3) argues this has shifted in the UK context at least to embrace a more openly class restructuring-type discourse where it is not just bad housing getting demolished, but a wholesale population and tenure restructure. Very little 'benefit' from urban regeneration flows to the poor at all, as this collection more than adequately shows. Instead, when inner city neighbourhoods are produced as 'deprived' or disadvantaged, they become the energetic focus of regeneration initiatives that are often unwanted and violently destructive of homes, social networks, cultural forms of life, and long-established mechanisms for economic survival.

Those most marginalized from the social and economic mainstream should not be housed in poor-quality buildings or environments, nor should they suffer due to lack of adequate attention from landlords and governments. Poverty in cities must be addressed. Yet poverty is a function of structural socio-economic inequalities. It seems obvious that the spatial fixes that urban regeneration tends to offer are not only useless in the face of this challenge, but in fact maintain the distribution of wealth and power in the direction of the already wealthy and powerful.

We find it astonishing that so little of the urban regeneration literature addresses the extent to which the interests of the development industry (and in this we include planners, designers, architects, developers, real estate agents, media, politicians, policy advisors, academics and others) are everywhere present in regeneration initiatives. Never is it reported in regeneration initiative 'outputs' how much money was made by landowners, developers and the vast array of consultants involved in regeneration. Nowhere is it analysed to what extent this industry delivers benefits to itself through the very act of drawing boundaries for 'area-based initiatives'. Baeten (2004) puts it brilliantly:

> The central notions of 'poverty' and 'deprivation' are repeatedly used in an uncritical manner by liberal urban analysts, who want to 'do good' for the city, but fail to acknowledge their paralysing effect on the emancipation of the urban disadvantaged. The use of these metaphors has, consequently, the opposite effect of what they were meant for. These commonly accepted metaphors for contemporary urban divisions ultimately lead to the disempowerment of the 'objects of study' (i.e. the poor, the excluded, the underclass), who become entangled in a permanent state of dependence on and subordination to charity, scrutiny and policy. In contrast, the 'knowing subjects' (i.e. academics, planners, politicians – mainly white, middle class, hetero-sexual and male), who magnanimously direct 'expertise, 'development plans' and 'budgets' toward 'deprived' urban quarters and their inhabitants, are further empowered.
>
> (p. 240)

What is more extraordinary is that none of this is really new. Criticisms of urban renewal policies in the United States during the 1950s and 1960s sounded similar warnings (see Jacobs 1961; Gans 1968). This makes our project in this book all the more crucial.

There are three contributions in this book that derive their intent from this 'deprivation' or 'exclusion' agenda. Two of them – evocative stories both – open and close the collection. One is a reflection from Doreen, a resident of the Silwood council housing estate in South London, and the other from Jessica, a resident of Silwood's neighbouring Pepys council housing estate. The chapter by Cruz on the revitalization of Green Bay, Wisconsin, also falls into this category, though it is not a state-imposed regeneration strategy.

As two of these contributions represent grass-roots action for change (Pepys and Green Bay), we want to treat them slightly differently. For us, these stories offer insights into the narratives of deprivation which are alive and well in regeneration rhetoric (in amongst the 'competitive city' noise). Yet they also offer small hopes for grass-roots initiatives where the material reality of people's lives, and the real challenges they face, can be transformed through genuinely regenerative work, where disinvestment and reinvestment are differently valued and understood. We return to these stories in the conclusion to this chapter, and to thinking differently about what reinvestment might mean in the final chapter.

Market-obeying strategies

Stories from Melbourne, Toronto, London and San Francisco show a qualitatively different suite of regeneration-inducing state activities from the quarter plan or the metropolitan-wide strategy for redevelopment. These are stories of gentrification, 'classic' style. Artists and progressives occupy the shabbier, downtrodden residences and warehouses of disinvested and deindustrialized city centres with the long-term working-class. Land values start to rise and eventually demand pushes prices beyond the reach of the original residents. The 'hotness' of property in San Francisco is better categorized super-gentrification where the ultra-rich are replacing the merely rich.

Quieter forms of urban policy are at work in these transformations: the sale of public land to the market in small lots; the development of financial incentives for property speculation through permissive tax regimes; the extension or provision of urban infrastructure seemingly unrelated to a particular development project; the granting of planning permission for redevelopment, site by site, block by block. While less overt, the strategy here is underpinned by the very same narratives of city competitiveness and cohesion as in those places subject to a specific urban policy intervention. Pro-development tendencies in 'city hall' are built on the expectation that investment better positions a city's economy in the global race for wealth creation, and that this creates better societies. The neoclassical theory of 'trickle-down' is alive and well in urban policy today.

Certainly wealth is created – the question is wealth for whom? As the stories here illustrate, in support of other research findings (Brewer *et al.* 2008), wealth is not created but becomes concentrated in fewer and fewer hands. Very rarely, and only in the most pathetic of trickles (e.g. underpaid, dead-end jobs serving the urban elite), does wealth disperse to those lower down the order of socioeconomic status. Residential displacement occurs through rising property prices and from the diversion of public money toward 'renaissance' projects, as Wyly and Hammel (2005:20) suggest. Permissiveness towards property markets results in displacement of those most marginalized from existing sources of wealth and power, predominantly because that permissiveness works to support property capital. As Cohen and Martí argue, in a 'growth-dependent economy ... development policy tends to formalize what "the market" wants to build' (this volume). Lehrer's story of Toronto (Chapter 15) is particularly telling. Here, the city government had a planning framework in place (and used it) to prevent development with inappropriate density and height, and lack of provision for additional infrastructure. Yet through protracted legal hearings and backroom negotiations, the proposals were eventually allowed. What the market wants to build actually *becomes* development policy.

This highlights a very important debate at the heart of this book: the role of policy in urban change. We are among those critics who claim that urban policy can be the creator, reinforcer and shaper of countless urban ills, as our commentary in this chapter shows. Yet we notice always the possibility held out by urban policy, the hope that it potentially offers to an equitable regeneration. Without a different kind of policy intervention, San Francisco's spiralling growth will continue to exclude low-income residents. In the absence of targeted regulation, Melbourne's music venues will continue to close. The BPP in Amsterdam creates space for subcultures that would otherwise have no place in that city. In Toronto, there was at least initial hope in the existence of planning frameworks and regulations to prevent unwanted and inappropriate development. In Berne's Lorraine district, housing associations enforce covenants on properties to prevent gentrification and promote communitarian housing forms.

None of this is easy, and certainly none of it comes without a fight. We note with interest that all the cities we define as having market-obeying strategies fall within Parts III and IV of this volume. All have been the sites of significant urban struggles and some have seen the invention of progressive, though far from perfect, urban policies in the face of aggressive gentrification. The question remains of whether a 'sweet spot' exists (Cohen and Martí this volume). Is it true that 'planners can really only win when they are channelling market forces into a more desirable direction, not where they try frontally to oppose them' (Gordon 2004:376)? We wonder, and return to these questions in the final chapter.

Conclusion: top down vs bottom up

This collection of stories suggests that there is greater potential for more equitable outcomes when there is grass-roots action in regeneration. It is neither foretold nor guaranteed, but the potential is worth exploring. The contributions to this book come from a range of perspectives, including several that actively supported a particular form of regeneration. Cruz's contribution (Chapter 11) from Green Bay, Wisconsin, tells of how the people of one neighbourhood in a small city made some tangible differences to their neighbourhood. Far from being a state-imposed regeneration initiative (though it was state supported), the residents and business owners on Broadway Street took action against what they perceived to be urban dereliction and social decay. The consequences are mixed, and Cruz offers a cautious view of the results so far. The process was not always consensual, and the politics surrounding notions of the 'deserving' and 'un-deserving' poor, and the growth of small business continue to bubble away under the surface. Parts of Green Bay are beginning to gentrify and Cruz notes how crucial the membership is of the On Broadway Inc. Board to keeping socially progressive options alive.

Regeneration of the Pepys estate in South London was also an entirely grass-roots-led strategy. As Jessica explains in her interview with Mark Saunders (this volume), the impetus came not from the physical dereliction of the estate, but from the loss of a suite of important socially oriented programmes. The funding that the Pepys Community Forum received was invested in social and community facilities such as a youth club and magazine, employment-related services, a credit union for low-income people, a community garden, and food cooperative. This lies in stark contrast to Doreen's experience of living through state-imposed physical regeneration on the Silwood social housing estate where displacement, fragmentation and loss of community facilities resulted.

The Green Bay and Pepys stories have two important elements in common: a group of local people working outside the interests of both the state and big property capital towards regeneration and a commitment to addressing social as well as physical issues that suggests a valuing of different kinds of 'reinvestment'. A commitment to socially progressive outcomes was also (at least partially) evident in the IBA Emscher Park in the Ruhr Valley. Müller and Carr take issue with many aspects of that regeneration process; however, they do point to some successes in the smaller regenerative projects such as the Riwetho housing community and the swimming pool-turned-social centre at Castrop-Rauxel. In Berne's Lorraine district, and in Amsterdam also, grass-roots movements are producing real alternatives with positive outcomes.

Incremental, partial approaches that accommodate a wide range of forms of reinvestment, beyond the fiscal, can work in the interests of genuine regeneration. Examples abound in this volume of the possibilities offered when reinvestment is de-coupled from its fiscal meaning. Reinvestment can mean the time and sweat of the local people who made a garden on a derelict site in Barcelona. It can mean the energy given by local people in forming housing cooperatives and protecting communal forms of housing occupation in Berne. It can mean the expertise and dedication of local people in the Pepys who have rebuilt the social infrastructure of their neighbourhood. All-encompassing strategies for change, by comparison, tend to quash alternative possibilities, by limiting the meaning of reinvestment to the financial contributions of property developers. We turn in the next chapter to the question of the politically contingent nature of regeneration, and look more carefully at the role of urban struggle, its potential, and at whether urban policy can play a more equitable and socially progressive role in the regeneration of cities.

24 Rising to a challenge

Kate Shaw

The purpose of this chapter is to discuss the role and potential of public policies that attempt to counteract the possible inequities of urban regeneration. Such policies can operate as state interventions that cut across market strategies or other government policies, working at cross-purposes to the urban regeneration agenda, or be embedded in existing policies as a modification of that agenda. We are largely sympathetic to Swyngedouw's (2007) assessment of policy in contemporary neoliberal society as having been 'evacuated of politics'; indeed we observe in Chapter 23 the extraordinary unanimity on urban regeneration policies in particular among planners, designers, architects, developers, real estate agents, media, politicians, policy advisors and some academics. Yet there are opportunities for dissent, critique and conflict ('politics', in short) where different policies are working towards different ends in the same place, for example. Opportunities exist also in policies that contain multiple objectives – whether inadvertently, as a result of vague planning and undefined intents (Cochrane 2007), or deliberately.

Amsterdam's *broedplaatsenbeleid* (breeding place policy) is a relatively simple case of the latter. The policy has various competing objectives: to provide low-cost spaces in the city for subcultural producers to live and work, to maintain the city's reputation for its alternative scene, to maintain the city brand, to encourage urban regeneration, to build a robust media and information technology industry, to legitimize reinvestment. Does the urban regeneration agenda contradict the continuity of low-cost living and working space? Yes, to the extent that urban regeneration is premised on and measured by rising land values. But the coincidence of interests between the scene and the brand creates an incentive for a more complex policy (see Shaw 2005b) in which state intervention changes the paradoxical nature of that relationship. To varying degrees, the breeding place policy has achieved all its objectives. Within the policy, between the various objectives and because of them, is a space for politics. A foundation set up by the activist group *de Vrije Ruimte* was able to develop and manage a live-work space according to the group's own terms, rather than those of the breeding place policy (though *de Vrije Ruimte*'s efforts have already made sure these terms are not that different). The activists made this further demand, and the city conceded it. Van de Geyn and Draaisma (this volume) point out that, '[f]or the city council, the building would become a temporary cultural hotspot. For the housing corporation, a cultural hotspot at the centre of its property holding meant an attractive creative environment to attract the new middle-class'. And, for low-income artists, activists and others, the former Volkskrant building means 'affordable rent and accessibility for all users'.

Policies working towards different ends in the same place also create room (and reason) for dissent. Melbourne is a clear instance of this. A mix of market-led strategies and state-driven urban regeneration policies was disrupted by two interventions in particular. One, in the inner municipality of St Kilda, fuelled by that coincidence of interests between the notion of the 'creative city' and the reality of an already existing creative culture, substantially modified a private-sector mega-project to enable the continuity of an important independent music venue. The second was the formation of a State taskforce to address the pressures of gentrification on the indie music scene city-wide. A set of policy interventions that would constrain the Australian national sport of property speculation was considered and rejected, with the bare minimum being implemented to keep the music scene alive. The very formation of the taskforce had an effect, however, and the interventions that were considered now sit in the public domain as an armoury of potential policies to be reassessed when gentrification pressures redouble. These actual and potential policy interventions, supported by a capital city arts strategy that starkly contradicts the city's approach to urban development, cut across existing urban regeneration strategies and policies. The policy-makers understand these tensions, and open up space for dissent by virtue of this obvious and acknowledged disjuncture.

In Barcelona and Salvador da Bahia, counteracting policies were embedded within urban regeneration policies – modifying the urban regeneration process to achieve different outcomes than originally intended. This is true at Barcelona's Can Ricart as well as in the Ciutat Villa: the masterplan for Poblenou was altered significantly from the original, though not to the satisfaction of the committed activists. In the gentrifying old town, the city's retail policy to ensure premises for low-income locals was a clear intervention into the market. It resulted in about 50 per cent of new shops meeting their needs, and 50 per cent servicing the desires of tourists and the new gentry. This can be read in the same way as the story of Salvador, with similar alternative interpretations. The urban regeneration strategies in both cities emphasize tourism, and we can presume that the city governments understand only too well that tourists want 'authenticity' in their holiday experience. It is a strategic and relatively simple matter to ensure some photogenic, long-term low-income residents in Barcelona, and (not too many) African-descended Brazilians to sell locally crafted souvenirs in the historic centre of Salvador. Yet Pascual-Molinas and Ribera-Fumaz (this volume) point out that the 50 per cent of premises oriented to the needs of poorer residents were delivered only because of neighbourhood opposition to gentrification, and that it can be seen as a change in city policy 'where meeting the needs of the local residents is a key objective'. Tarsi (this volume) argues that it was popular struggle that produced the integration of the urban regeneration policy 'with a social policy allowing old residents to remain in the Pelourinho, affirming the right of the poor to the city'. The existence of these modifying policies, oriented on some level towards greater social equity in the urban regeneration process, opens up space for further demands.

As in Amsterdam, both (and more) interpretations are probably true. None of these stories are unqualified. They are full of half-full and half-empty glasses, babies and bathwaters, buts, howevers and on the other hands. What is absolutely clear in every instance is the role of struggle in bringing policy responses from these cities this far. The last two stories in this book are to us most instructive. With Amsterdam, they have produced levels of government response which, in their contexts, are as equitable, for whatever motives, as anything we have seen.

Notwithstanding San Francisco's state of super-gentrification with seemingly infinite global demand for space, local planning and housing activists – building on a long and venerable tradition of activism in that city – have succeeded in bringing about policy interventions into housing and service provision that other cities with perhaps more incentive still baulk at. Unlike Amsterdam, Barcelona, Salvador da Bahia and Melbourne, there is little to be gained on the face of it by San Francisco maintaining places for low-income people, whether 'creative' or not. Of course, neither does that city apparently need more regeneration (or further gentrification), which is largely market driven. Three key policy initiatives were driven by grass-roots struggle and implemented by government: strong inclusionary housing laws, policies for public benefits zoning, and a strong city resolution oriented to future equitable development. They have succeeded in reducing displacement and providing decent housing and services for a significant number of San Francisco's lower income residents.

Cohen and Martí (this volume) ask, 'Are we trapping ourselves by thinking that we can "manage" gentrification?' They put the dilemma beautifully:

> A handful of progressives are questioning the entire notion of trying to find the 'sweet spot' between the interests of private sector gentrification-inducing development and preservation of an equitable, diverse, politically progressive city. 'Public benefits' depends on supporting capital, bringing in investment (which these days means extremely fluid global finance) to pay directly and indirectly for public needs. Can a strategy of mutual accommodation ever match the growth machine or create 'equitable' development policies?

What do they conclude? Tightening the screws harder; strengthening the role of the state. Trying to set firmer limits on the operations of the market, emphasizing the 'equitable' over the 'development'. In an approach that suggests Friedmann's (2007) argument for 'endogenous' development of city regions, in which real investment is made in local assets rather than relying on exports (bringing in capital from outside), Cohen and Martí argue for 'treating the city and its people as the assets, with development given the *privilege* of investing in its growth and improvement on terms more explicitly spelled out by and for the city's communities'.

We have observed elsewhere the irony that gentrification proceeds most confidently in the places that need new investment least (Shaw 2005a), and perhaps here is the incentive for the City of San Francisco to maintain these interventions: it does not have much to lose. Investors and developers do threaten that 'development may come to a grinding halt in San Francisco' if the screws are tightened too hard, but the city's progressive advocates are tough. So far, it appears, no party has folded, and Cohen and Martí's story suggests the progressives are gaining. That threat of a 'grinding halt to development' is heard all over the world, of course, and most governments capitulate – especially in those cities that are heavily disinvested and would really benefit from reinvestment. But San Francisco gives cause for reflection. The investors and developers have not walked away. For those other cities, we return to our original questions: can the benefits of reinvestment be harnessed without excluding vulnerable residents? Are such approaches politically viable in the long term?

Part of an answer to these questions is found in the city of Berne in the district of Lorraine, where a period of disinvestment was followed by plans for urban regeneration that did not happen in the usual way. For various reasons, including the global recession in the 1970s, growing environmental awareness throughout Europe, and a culture of radical activism in Switzerland that flourished relatively late (in the early 1980s), most of the large regeneration projects never eventuated. But some regeneration did occur, and Stienen and Blumer (this volume) observe that the city's *alternativszene* did not resist it. Instead, the activists 'tried to shape it', and were instrumental in producing policies that increased the social housing stock and strengthened security of tenure in squats and private rental. They were able to ensure that new construction 'adapted to the particular conditions of the neighbourhood's residents' and allowed 'the individual needs, preferences and economic possibilities' of local residents to diversify. The authors of this chapter appear to agree with the municipality's assessment of the neighbourhood as displaying '"a high quality of life, a diversified housing market which meets a variety of social needs", and "a rich multicultural life"'. No people were displaced in the process.

What kind of policies are these? They are substantial modifications to an already existing urban regeneration agenda. The authors of the story of Berne, activists themselves, call the process in which they participated 'politically negotiated regeneration'. This is regeneration without negative effects (the one disturbing development that the authors do refer to – the socio-cultural marginalization of the conservative working-class, which is perhaps becoming more politically extreme as a consequence – is a fascinating nuance but cannot be attributed solely to the regeneration process). The district of Lorraine went through an equitable form of urban regeneration without displacement.

All these stories tell us that policy that counters the potential inequities of urban regeneration – inequities which *will* occur in the absence of intervention – is possible. We know that any and every element of equitable policy in urban regeneration has to be fought for. Can these stories help us frame a new approach to this 'urban renaissance' that so many cities all over the world are seeking to embrace?

Towards a radical approach to reinvestment

We assume that processes of disinvestment and reinvestment are 'an inherent and integral part of the deterritorialized and reterritorialized relations of global neoliberal capitalism' (Swyngedouw 2007:31), although in their basic form they pre-date capitalism. Urban regeneration, as a process of reinvestment in a place after a period of disinvestment, is a vital element of capitalist relations – though it too can be separated from that paradigm. Smith (2002a) argues that the long-term solution to urban regeneration that causes displacement – that is, gentrifica-tion – is to circumscribe the incentives for sustained disinvestment in the first place. Of course he is right, and this requires breaking down the relations of global neoliberal capitalism. We are not sure this is going to happen soon. In the meantime, we want a radical approach to reinvestment that satisfies the anxious need of governments of disinvested cities for jobs, activity and infrastructure improvements, and delivers humane and equitable results for their people.

We have shown that urban regeneration can be understood as a process that need not lead to displacement, and does not by definition have the class charac-ter that is inherent in gentrification. There is a possibility, then, for the process to carry out a more radical form of reinvestment: one that delivers secure and affordable housing and decent services and jobs to people on low incomes. To explore this, we need to rethink the processes of dis- and reinvestment. We are reminded that,

> [u]rban and regional development is not a smooth process towards an imag-inary equilibrium state. Instead … it lurches forward from imbalance to imbalance, as different pressure points become activated, forcing an adjust-ment in policies and budgetary allocations.
>
> (Friedmann 2007:996)

No state is stable, and for as long as social, economic, cultural, environmental and political conditions shift, there will be fluctuations in relative levels of disinvest-ment or reinvestment. It appears that the greater these fluctuations, the more vio-lence is done to the people who experience them. Those least protected by wealth remain most vulnerable to massive withdrawals of capital from an area, and to rapid reinvestment. Minimal fluctuations over longer periods of time cause less damage. It is possible to regard a certain level of fluctuation as desirable. Consider the uses to which disinvested spaces are put: often they become spaces of resistance. There is a certain circularity to the argument that one positive of global neoliberal capitalism is that it produces spaces in which a politics of resist-ance to global neoliberal capitalism can form, indeed. But such spaces may be desirable even in a post-capitalist utopia.

In Istanbul and Salvador, disinvested spaces are meeting places for radical thought and political organization. Disinvested spaces create opportunities for a range of alter-native practices, including music, poetry, theatre, street art and *detournement*. Surplus housing and working spaces are valuable precisely because no one wants them: their 'loose fit' can accommodate uses and activities that no one has thought of yet. They allow and encourage imagination. Disinvestment creates cheap space, which can be lived and worked in and otherwise occupied by people who through circumstance or choice have low incomes and few options. Low-cost space pro-vides security and allows people to take risks.

This book contains stories from Berne, Amsterdam, Melbourne, Barcelona, Toronto, London, Birmingham, Beijing and Berlin, of people making good use of disinvested city space. Low-income housing made it possible for people in Berne, Toronto and London 'to develop unconventional and alternative forms of urban life' (Stienen and Blumer, this volume). Vacant warehouses in Amsterdam and Melbourne were squatted or rented cheaply, providing 'not only artist spaces, but spaces for craftsmen, activists and other creative producers' (van de Geyn and Draaisma, this volume). Deindustrialized factories in Barcelona and Birmingham were converted to vital cultural and community centres. Beijing's old town, for so long forgotten by the rest of the city, became a hub of bohemian dissent. In Berlin, the tag 'poor but sexy' (Lindner 2007) is not just an account of its compelling cultural status, but an explanation for it.

Now it can be argued that these alternative occupations in themselves constitute a process of reinvestment, and so they do, in their use value. There is a well-worn theme in the gentrification literature that proposes that squatters and artists are the 'pioneers' of gentrification (Zukin 1982; Caulfield 1994; Ley 1996). It ranges from a benign analysis in which they are unwitting, and ultimately displaced (Ley 1996; Rose 1996), to a more trenchant critique that says they are knowing participants in gentrification (Treanor 2002). Smith (1996) argues conversely that these 'marginal gentrifiers' are indeed marginal to the process in as much as they are defined by their low incomes and gentrification is defined by its transition to more affluent users. The critical point here is that reinvestment in use value is a qualitatively and quantitatively different process to reinvestment in exchange value, or revalorization.

A radical approach to reinvestment emphasizes use over exchange. It values disinvested space; it upgrades infrastructure slowly. It values low-income people and orients urban regeneration to their advantage. It puts mechanisms in place to ensure that people are not displaced. *Behutsame Stadterneuerung* – the 'cautious urban renewal' process in Berlin – had a clear focus on reinvesting in the city after its drastic disinvestment. Bader and Bialluch (this volume) describe the process as follows:

> Behutsame Stadterneuerung adopted mechanisms such as rent controls …
> and strong tenant participation to bring about 'urban renaissance'. The main
> aims of this paradigmatic policy were participation, preservation of the spe-
> cific character of neighbourhoods and the old building stock, neighbourhood
> improvement according to the needs of the inhabitants, gradual renewal of
> buildings, and solid financial support for this policy (Hämer 2007).

As in Berne and Amsterdam, residents and users of the places being regenerated were involved in the process, and not displaced. In Berlin, the recent shift to a 'classic' urban regeneration policy has still not led to displacement, though Bader and Bialluch argue that this is due more to the weak local economy than any clear social justice objectives of the local state. More telling is the grass-roots opposition to this shift, which the authors say 'has already generated strong pressure on the local government'. It was struggle in Berlin that delivered the 'cautious urban renewal' approach in the first place, and this story can yet take many turns.

How useful is this approach to reinvestment? For still relatively disinvested places, like the Ruhr Valley and Leipzig, the conversation between all the cities in this volume has profound implications. The cities in Parts III and IV tell those in Part II that they will not remain in this disinvested state. The population of the world is growing exponentially, finance and property capital is increasingly foot-loose – it is just a matter of time before all disinvested places experience some sort of reinvestment. The weak economies in the Ruhr Valley and Leipzig hold potential for a different approach, in which protective infrastructure against gentrification can be put in place now, while it is still very possible.

This approach requires reconsideration and perhaps redefinition of what are 'problems' and 'solutions'. Currently the unsuccessful processes of regeneration in Leipzig and the Ruhr Valley are seen as problematic. Some reinvestment has occurred but the regeneration is inhibited, rents have not risen, and the discussion is about what to do with the surplus housing stock. But others are already questioning this approach. Voss (2007), for example, argues in the context of the Ruhr that the surplus housing

> might relieve the daily stress faced by children of people on social welfare or low-level unemployment benefit. ... It stands to reason that market forces should also apply when they are unfavourable to landlords, and lead to rents that young entrepreneurs, artists, and other creative people or, in a word, any-one with lots of ideas but little money, can afford. Furthermore, the subse-quent devaluation of real estate could lead groups that previously had no chance of buying to become owner-occupiers.
>
> (p. 8)

Taking advantage of a possibly quite temporary phase of disinvestment with an increase in home ownership might be problematic if the purchasers require mort-gage credit. But state intervention and/or assistance in this context can take many forms. Perhaps more conservatively, this is an opportune moment for the state to invest in social housing, where long-term secure rental for the people Voss discusses can be at least as satisfactory.

It is interesting to reflect on the different perspectives and expectations of the con-tributors to this volume. The accounts of the state in Part I of this book range from violently oppressive (Istanbul and Johannesburg) to benign but misguided (Mexico City). Strong critiques of the state come also from Amsterdam and Berlin (and indeed Berne, in the final part of the story), as one might expect. The gulf between the actualities of these governments, though, is enormous. The authors of the Part I stories would likely see the interventions of those governments in Part IV as com-plex, light-handed and genuinely approaching a radical form of reinvestment. Of course, we realize that one person's success can be another's disappointment. What we find inspiring is the relentlessness of the critique and the persistence of the oppo-sition. No state is stable, nor does the desire to make a better world ever rest.

The key is continuing struggle. Eric Clark proposed in 2005 that a comparative analysis aimed at understanding why this process turns into tumult in some places and not others would find two key factors to be degree of social polarisation and practices surrounding property rights. In places characterised by a high degree of social polarisation, short on legally practised recognition of the rights of users of place and long on legally practised recognition of the rights of owners of space, the conflict inherent in gentrification becomes inflammatory. Not so in places characterised by relative equality and legally practised recognition of the rights of users of place. If so, this indicates a direction for political engagement aimed to curb the occurrence of gentrification and to change societal relations such that when it does occur (and it will), conditions are established for more benign ends.

(p. 262)

In Istanbul, Johannesburg, Riyadh, Mexico City and Salvador, at least, Clark's analysis is clearly accurate. In these cities, social polarization is high and the exchange value of space is much more highly rated than use value. But then there is the conflict that results from a confident opposition, found in cities exhibiting precisely the opposite characteristics: the social democracies of Amsterdam, Berne and Berlin. This is a different kind of violence, pushing the boundaries, taking the 'game', as Cohen and Martí refer to the less physical but just as tense conflict in San Francisco, to new levels.

A radical approach to reinvestment must be continually fought for. It requires interventions to reduce social polarization and policies that emphasize use value. Its political viability depends on forever making the case, protecting gains, pushing for more. It requires progressive politicians, policy-makers and planners, with support from activists and academics and relevant research. It takes a long time, and the battle never ends.

Clark (2005) argues that,

[t]his kind of comparative analysis is strikingly absent in the gentrification literature. Academia, it seems, does not encourage interest in policy issues and political engagement, rewarding instead awareness of the 'chaos and complexity'. While there is no lack of critique of gentrification as a strategic policy, there is a dearth of effort to outline alternatives or to the variability of grounded impacts in a wider variety of settings. This poses a considerable challenge to gentrification research.

(p. 262)

We hope this book rises to that challenge.

Bibliography

16 Hoxton Square (2007) *Homepage*. Online. Available HTTP: http://www.16hoxton square.com (accessed 27 August 2007).

AA.VV. (ed.) (2007) *Modello Roma: L'ambigua modernità*, Rome: Odradek Edizioni.

ABS (2006) 'Australian Bureau of Statistics census data, 2006', Canberra: ABS.

Abteilung Zukunft (2007) *Reitschule Bern: 20 Jahre und mehr*, Zürich: Edition 8.

Active 18 (2006) *Queen West Triangle Charrette: Proceedings*. Online. Available HTTP: http://www.active18.org/charrette.html (accessed 10 October 2007).

AEP (2003) 'T.C. 58. Hukumet Acil Eylem Planı' (Emergency Action Plan of 58th Government of Turkish Republic), Ankara: Basbakanlık.

Agenzia Regionale per la Protezione Ambientale della Toscana (2007) 'III Rapporto Sulla Qualità dell'Ambiente Urbano', Firenze: ARPAT.

Adjuntament de Barcelona (2000) 'Modificació del PGM per les Àrees Industrials del Poblenou: Districte d'Activitats 22@', Barcelona: Sector d'Urbanisme. Ajuntament de Barcelona.

—— (2006a) 'Modificació del Pla de Millora Urbana de la UA – 1 del peri del secor el parc central. Preservació del Recinte de Can Ricart', Barcelona: Districte d'activitats 22@bcn Ajuntament de Barcelona.

—— (2006b) 'Modificació del Pla Especial Arquitectònic i Historicoartístic de la Ciutat de Barcelona. Districte de Sant Martí. Patrimoni industrial del Poblenou', Barcelona: Ajuntament de Barcelona.

—— (2007) *Anuari Estadísitc de la Ciutat de Barcelona 2007*, Barcelona: Ajuntament de Barcelona. Online. Available HTTP: http://www.bcn.cat/estadistica/catala/dades/anuari/index.htm (accessed 23 May 2008).

Allegretti, G. (2006) 'Lavorare sulle energie di contraddizione: esperienze di coinvolgimento degli abitanti nel Comune di Roma', in G. Allegretti and E. Frascaroli (eds) *Percorsi condivisi*, Florence: Alinea.

—— (2008) 'Under a peculiar sky: Recent trends of dialogic public policies in the territory of Rome', in E. Braathen and D. Chavez (eds) *Progressive Cities*, Amsterdam: TNI.

Allen, C. (2008) *Housing Market Renewal and Social Class*, New York: Routledge.

Allen, C. and Crookes, L. (2008) 'Fables of the reconstruction: A critical phenomenology of "place shaping" in the North of England', paper presented at Regional Studies Association Conference, University College London, March 2008.

Amado, J. (1977) *Bahia de Todos os Santos: Guia de ruas e mistérios*, Rio de Janeiro: Record.

—— (1999) *Sudore*, Torino: Einaudi.

Amin, A., Massey, D. and Thrift, N. (2000) *Cities for the Many Not the Few*, Bristol: Policy Press.

Amnesty International (2007) *People's Republic of China: The Olympics countdown – one year left to fulfil human rights promises, ASA 17/024/2007.* Online. Available HTTP: http://www.amnesty.org/en/library/info/ASA17/024/2007/en (accessed 26 April 2008).

Ar-Riyadh Development Authority (ADA) (2004) 'Riyadh strategic plan: Local structure plans volumes 10–13', Riyadh: Ar-Riyadh Development Authority.

—— (2005) 'Survey of population and uses in the city of Riyadh', Riyadh: Ar-Riyadh Development Authority, unpublished report.

Asamblea Legislativa del Distrito Federal 'Ley de Desarrollo Urbano del Distrito Federal' (1996) *Gaceta Oficial del Distrito Federal, 29 January.* Online. Available HTTP: http://www.asambleadf.gob.mx/al/pdf/010803000011.pdf (accessed 28 January 2008).

Associació Veïns Poblenou (2001) 'Habitatges afectats pel 22@: no és això, veïns, no és això', *El Poblenou*, 26: 6.

—— (2002a) 'Crisi al 22@ o benefici del desconcert', *El Poblenou*, 27: 6–8.

—— (2002b) 'Per un 22@ amb participació i sense especulació', *El Poblenou*, 28: 9–12.

Atkinson, Rob (2000) 'Combating social exclusion in Europe: The new urban policy challenge', *Urban Studies*, 37(5/6): 1037–55.

Atkinson, Rowland (2000a) 'The hidden costs of gentrification: Displacement in central London', *Journal of Housing and the Built Environment*, 15(4): 307–26.

—— (2000b) 'Professionalisation and displacement in Greater London', *Area*, 32(3): 287–95.

—— (2002) 'Does gentrification help or harm urban neighbourhoods?: An assessment of the evidence-base in the context of the new urban agenda', ESRC Centre for Neighbourhood Research Paper No. 5.

—— (2003) 'Domestication by cappuccino or a revenge on urban space? Control and empowerment in the management of public spaces', *Urban Studies*, 40(9): 1829–43.

—— (2004) 'The evidence on the impact of gentrification: New lessons for the urban renaissance?', *European Journal of Housing Policy*, 4(1): 107–31.

Atkinson, R. and Bridge, G. (2005) 'Introduction: The new urban colonialism', in R. Atkinson and G. Bridge (eds) *Gentrification in a Global Context: The New Urban Colonialism*, London: Routledge.

Atkinson, R. and Easthope, H. (2007) 'The consequences of the creative class: The pursuit of creativity strategies in Australia's cities', paper delivered to the State of Australian Cities national conference, Adelaide, 28–30 November.

Bade, F.-J. (2006) *Strukturwandel und Wettbewerbsfähigkeit der Regionen NRWs.* Online. Available HTTP: http://www.raumplanung.uni-dortmund.de/rwp/20060929/TxtPublik.htm (accessed 20 October 2007).

Bader, I. (2004) 'Subculture – pioneer for reindustrialization by the music industry or counterculture?', in INURA (ed.) *Contested Metropolis: Six Cities at the Beginning of the 21st Century*, Basel: Birkhäuser.

Badyina, A. and Golubchikov, O. (2005) 'Gentrification in central Moscow – a market process or deliberate policy?: Money, power and people in housing regeneration in Ostozhenka', *Geografiska Annaler B*, 87(2):113–29.

Baeten, G. (2000) 'From community planning to partnership planning: Urban regeneration and shifting power geometries on the South Bank, London', *GeoJournal*, 51(4): 293–300.

—— (2004) 'Inner-city misery: Real and imagined', *City*, 8(2): 235–41.

Balibrea, M.P. (2004) 'Urbanism, culture and the post-industrial city: Challenging the "Barcelona Model"', in T. Marshall (ed.) *Transforming Barcelona*, London: Routledge.

Barker, K. (2004) *Review of Housing Supply*, London: HM Treasury.

Barreto, L.M. and dos Santos, M.A.S.C. (2002) 'Pelourinho: Desenvolvimento socioeconomico', Salvador: Secretaria da Cultura e Turismo.

Bayraktar, E. (2006) 'Gecekondu ve Kentse el Yenileme' (Sqautter Settlement and Urban Renewal), Ankara : Ekonomik Arastırmalar Merkezi Yayinlari.

Beauregard, R. (2004) 'Are cities resurgent?', *City*, 8(3): 421–27.

Beck, C. (2002) 'Die Erwartungen der Wohnungswirtschaft', in Deutscher Verband für Wohnungswesen (ed.) *Städtebau und Raumordnung e.V. 2002: Wohnungswirtschaftlicher Strukturwandel in Ostdeutschland. Redebeiträge zum Schwerpunktthema des Deutschen Verbandes für Wohnungswesen, Städtebau und Raumordnung e.V. (Tagungsdokumentation)*, Berlin.

BeLa Partnership (2007) Priors Hall Housing Strategy, Corby: Corby Borough Council.

Bénit-Gbaffou, C. (2006) 'In the shadow of the 2010 Greater Ellis Park development: Decision-making and the poor in inner city Johannesburg', paper presented at the Life of the City Conference, University of the Witwatersrand, Johannesburg, 4–6 September.

Benlliure, P. (2006) 'Las dos caras de la expansión urbana en la ZMVM: debate sobre el reclamiento vs. "el desbordamiento urbano"', paper presented at Universidad Autónoma Metropolitana-Azcapotzalco Mexico City Debate Programme, Mexico City, 23–25 February.

—— (2007) 'Situación de la vivienda en la ZMVM 2000–2006. Efectos del Bando 2 y perspectivas post-bando', paper presented at Universidad Iberoamericana 'Diplomado en Vivienda' Programme, Mexico City, October.

Bentley-Mays, J. (2006) 'The next step in remaking Queen West: Activists, architects and developers need to start a conversation', *Globe and Mail*, 17 March.

Berger, J. (1987) *Kreuzberger Wanderbuch: Wege ins widerborstige Berlin*, 2nd edn, Berlin: Goebel.

Bernt, M. (2003) *Rübergeklappt: Die "behutsame Stadterneuerung" im Berlin der 90er Jahre*, Berlin: Schelzky & Jeep.

—— (2005) 'Demolition program east', in P. Oswalt (ed.) *Shrinking Cities – Volume 1: International Research*, Ostfelden-Ruit: Hatje Cant Publishers.

Bernt, M. and Holm, A. (2002). 'Gentrification in Ostdeutschland: der Fall Prenzlauer Berg', in *Deutsche Zeitschrift für Kommunalwissenschaften (DfK)*, 41: 125–50.

Bianchini, F. and Parkinson, M. (eds) (1993) *Cultural Policy and Urban Regeneration*, Manchester: Manchester University Press.

Birmingham City Council (1996) 'Digbeth Millennium Quarter Plan', Birmingham: Birmingham City Council.

—— (2001) 'Eastside Development Framework', Birmingham: Birmingham City Council.

—— (2002) 'City Centre Canal Corridor: Development Framework', Birmingham: Birmingham City Council.

—— (2006) 'The Birmingham City Council (Curzon Street etc City Centre) Compulsory Purchase Order 2006: Statement of Reasons for Making the Order', Birmingham, Birmingham City Council.

Blomley, N. (2004) *Unsettling the City: Urban Land and the Politics of Property*, New York: Routledge.

Blumer, D. and Tschannen, P. (2001) '"Wer hat das Sagen im Quartier?" Einflussmöglichkeiten von Akteurgruppen auf die Entwicklung zweier Quartiere der Stadt Bern (Breitenrain und Lorraine)'. Bern: Institut für Soziologie, Universität Bern.

—— (2006) 'Machtkampf ums Quartier–das Berner, Nordquartier', in A. Stienen (ed.) *Integrationsmaschine Stadt?: Interkulturelle Beziehungsdynamiken am Beispiel von Bern*, Bern: Haupt.

BMVBS (2007) *Veränderung der Anbieterstruktur im deutschen Wohnungsmarkt und woh-nungspolitische Implikationen*. Forschung vol. 124. ed. Bundesministerium für Verkehr, Bau und Stadtentwicklung, Bonn: Bundesamt für Bauwesen und Raumordnung.

Boeckmann, K. (2006) *Ende einer Talfahrt?: Entwicklung der Beschäftigung im Oestlichen Ruhrgebiet*, Dortmund: Sozialforschungsstelle Dortmund.

Boehme, G. (1995) *Atmosphäre*, Frankfurt a.M.: Suhrkamp.

Boelhouwer, P. (2002), 'Trends in dutch housing policy and the shifting position of the social rented sector', *Urban Studies*, 39(2): 219–35.

Boemer, H. (1999) 'Moeglichkeiten und Grenzen des regionalpolitischen und struktur-politischen Beitrags der IBA', in S. Mueller and R. Herrmann (eds) *Inszenierter Fortschritt. Die Emscherregion und ihre Bauausstellung*, Bielefeld: Verein zur Förderung alternativer Kommunalpolitik AKP.

Bonet-Martí, J. (2004) 'Barcelona: la reinvención de la ciudad portuaria en la nueva economía global', *Archipiélago*, 62: 1–5.

Bourdieu, P. and Wacquant, L. (2001) 'Neoliberal newspeak: Notes on the new planetary vulgate', *Radical Philosophy*, 105 (January–February): 2–5.

Brasserie Lorraine (2001) Denn wir gehen nicht unter: 20 Jahre Genossenschaft Brasserie Lorraine Bern: ein Rückblick mit Ausblick, Bern: Genossenschaft KuKuZ.

Breek, P. and de Graad, F. (2001) *Laat duizend vrijplaatsen bloeien: Onderzoek naar vri-jplaatsen in Amsterdam*, Amsterdam: De Vrije Ruimte.

Brenner, N. and Theodore, N. (eds) (2002a) *Spaces of Neoliberalism: Urban Restructuring in North America and Western Europe*, Oxford: Blackwell.

—— (2002b) 'Cities and the geographies of "actually existing neoliberalism"', *Antipode*, 34(3): 349–79.

—— (2005) 'Neolibralism and the urban condition', *City*, 9(1): 101–07.

Brewer, M., Muriel, A., Phillips, D. and Sibieta, L. (2008) *Poverty and Inequality in the UK 2008*, Institute for Fiscal Studies. Online. Available HTTP: http://www.ifs.org.uk/publications.php?publication_id=4258 (accessed 11 June 2008).

Bridge, G. and Dowling, R. (2001) 'Microgeographies of retailing and gentrification', *Australian Geographer*, 32(1): 93–107.

Brindley, T. (2000) 'Community roles in urban regeneration: New partnerships on London's South Bank', *City*, 4(3): 363–77.

British Property Federation (2004) 'Manifesto', London: British Property Federation.

Buğra, A. (1998) 'The immoral economy of housing in Turkey', *International Journal of Urban and Regional Research*, 22(2): 303–17.

Büro Herwarth and Holz (2006) Voruntersuchung/Machbarkeitsstudie. Kreuzberg – Spreeufer, Berlin 2005, Berlin: Senatsverwaltung für Stadtentwicklung [in cooperation with Bezirksamt Friedrichshain-Kreuzberg].

Butler, T. (2003) *London Calling: The Middle Classes and the Re-making of Inner London*, Oxford: Berg.

Calavita, N. and Ferrer, A. (2000) 'Behind Barcelona's success story: Citizens movements and planners' power', *Journal of Urban History*, 26(6): 793–807.

Cameron, S. (2006) 'From low demand to rising aspirations: Housing market renewal within regional and neighbourhood regeneration policy', *Housing Studies*, 21(1): 3–16.

Campbell, C. (2008) *WQW: Active 18 in Planning Land 2007 Summary Report*. Online. Available HTTP: http://active18.org/releases/report-2007-summary/WQW_2007_Summary. pdf/view (accessed 18 February 2008).

Cantarino, C. (2007) *Monumenta muda pelos moradores dos centros*. Online. Available HTTP: http://www.revista.iphan.gov.br/materia.php?id=103 (accessed 10 January 2008).

Capel, H. (2005) *El Modelo Barcelona: un Análisis Crítico*, Barcelona: Ediciones Del Serbal.

Cardoso P. and Saule, N.J. (2005) 'O Direito à Moradia no Brasil – Violações, Práticas Positivas e Recomendações ao Governo Brasileiro', São Paulo: Instituto Pólis.

Carley, M. and Kirk, K. (1998) *Sustainable by 2020? A Strategic Approach to Urban Regeneration for Britain's Cities*, Bristol: The Policy Press.

Carmona, M. (2001) 'Implementing urban renaissance: Problems, possibilities and plans in South East England', *Progress in Planning*, 56(4): 169–250.

Carpenter, J. and Lees, L. (1995) 'Gentrification in New York, London and Paris: An international comparison', *International Journal of Urban and Regional Research*, 19(2): 286–303.

Cartner-Morley, J. (2003) 'Where have all the cool people gone?', *The Guardian*, 21 November. Online. Available HTTP: http://arts.guardian.co.uk/features/story/0,,109 0073,00.html (accessed 27 August 2007).

Caudo, G. (2007) 'La questione abitativa e le nuove forme dell'abitare', in AA.VV. (ed.) *Modello Roma. L'ambigua modernità*, Rome: Odradek Edizioni.

Caulfield, J. (1994) *City Form and Everyday Life: Toronto's Gentrification and Critical Social Practice*, Canada: University of Toronto Press.

Cellamare, C. (2007) 'Le insidie della partecipazione', in AA.VV. (ed.) *Modello Roma: L'ambigua modernità*, Rome: Odradek Edizioni.

Centre on Housing Rights and Evictions (COHRE) (2004) 'Any room for the poor?', unpublished draft report, Johannesburg: Centre on Housing Rights and Evictions.

China Daily (2007) Save our hutong, *China Daily,* 16 May: 10.

City Fringe Partnership (2003) *Unemployment in the City Fringe Area of London: Causes and Recommendations for Action*, London: Corporation of London. Online. Available HTTP: http://www.cityfringe.org.uk/cms/images/12%20BC_RS_uncityfringe_0304_FR.pdf (accessed 27 August 2007).

City of Green Bay (2003) *Green Bay Smart Growth 2022 Comprehensive Plan*. Online. Available HTTP: http://www.ci.green-bay.wi.us/geninfo/planning_development/planning/planning_smartgrowth_o.html (accessed 4 June 2008).

City of Johannesburg (CoJ) (2002) 'Jo'burg 2030 vision', Johannesburg: City of Johannesburg.

—— (2003a) 'Inner city regeneration strategy', Johannesburg: City of Johannesburg.

—— (2003b) 'Regional spatial development framework', Johannesburg: City of Johannesburg.

—— (2006a) 'Growth and development strategy', Johannesburg: City of Johannesburg.

—— (2006b) 'Human development strategy', Johannesburg: City of Johannesburg.

—— (2007) 'Inner city regeneration charter', Johannesburg: City of Johannesburg.

City of Melbourne (1999) 'Melbourne, a city for the arts, City of Melbourne's Cultural Policy'. Online. Available HTTP: http://www.melbourne.vic.gov.au/upload/CPSummary.pdf (accessed January 2003).

—— (2002) 'Municipal strategic statement', Melbourne: City of Melbourne.

—— (2004) 'Arts strategy 2004–2007', Melbourne: City of Melbourne.

—— (2007a) 'Analysis of population and housing, 2001–2006', Melbourne: City of Melbourne.

—— (2007b) 'Housing the arts in the City of Melbourne', report to Council, September.

City of Melbourne and Jan Gehl Architects (2004) 'Places for people', Melbourne: City of Melbourne.

City of Toronto (1976) 'Trends and planning goals South Parkdale', Toronto: City of Toronto Planning Board, April.

—— (2002) 'Toronto official plan', City Planning Division, Toronto: City of Toronto.

—— (2003) 'Toronto's culture plan for the creative city', Toronto: City of Toronto Culture Division.

—— (2005a) 'Toronto staff report: Preliminary report, rezoning, 1171 Queen Street', Toronto: City of Toronto, 14 June.

—— (2005b) 'Toronto staff report: Further report, 48 Abell', Toronto: City of Toronto, 1 September.

—— (2005c) 'Planning department staff report', Toronto: City of Toronto, 9 November.

—— (2005d) 'Staff report: Request for zoning review for West Queen West Triangle area'. Online. Available HTTP: http://www.Toronto.ca/legdocs/2005/agendas/committees/te/te051115/it081.pdf (accessed 14 March 2008).

—— (2006) 'Staff report: Preliminary report, rezoning application 150 Sudbury', Toronto: City of Toronto, 23 January.

Clark, E. (2005) 'The order and simplicity of gentrification – a political challenge', in R. Atkinson and G. Bridge (eds) *Gentrification in a Global Context: The New Urban Colonialism*, London: Routledge.

Cochrane, A. (2007) *Understanding Urban Policy: A Critical Approach*, Oxford: Blackwell.

Cohen, N. (2007) 'Why the bogey woman who wants to build over the green belt might just be creating a country worth living in', *New Statesman*, 30 July: 26–28.

Colenutt, B., Rhodes, M. and Stevens, G. (2007) 'Meeting the challenge of the growth agenda: A learning region for North Northants', Northampton: University of Northampton.

College Amsterdam (2006) 'Amsterdam Topstad: Metropool, Amsterdam terug in de top 5 van Europese vestingslocaties', Amsterdam: Gemeente Amsterdam.

Colomb, C. (2007) 'Unpacking New Labour's "Urban Renaissance" agenda: Towards a socially sustainable reurbanization of British cities?', *Planning Practice and Research*, 22(1): 1–24.

Commonwealth of Australia (2004) Productivity Commission Inquiry Report, No. 28, March.

Comune di Firenze (2006) *Piano di gestione del Centro Storico 2006–2008*. Online. Available HTTP: http://www.comune.firenze.it/unesco/piano_gestione.html (accessed 3 July 2007).

—— (2007a) *Piano strutturale: Progetto, relazione generale, versione adottata*. Online. Available HTTP: http://www.comune.firenze.it/opencms/export/sites/retecivica/materiali/dir_urbanistica/P_RelazioneGenerale.pdf (accessed 10 August 2007).

—— (2007b) *Piano strutturale, versione adottata*. Online. Available HTTP: http://www.comune.firenze.it/opencms/export/sites/retecivica/materiali/dir_urbanistica/P_Relazione Generale.pdf (accessed 10 September 2007).

CONDER (1996) 'Estudo de circulaçao e transporte', Salvador: Unesco-Conder-Governo da Bahia.

Councillor Cowan (2005) 'Slumbering giant wakes and flexes its muscles: The CBD is poised for a revival as years of planning and action start to bear fruit', *The Sunday Times*, 6 November.

Cowans, J., Robinson, D., and Meikle, J. (2007) 'Large scale housing growth in North Northamptonshire: Challenges and opportunities', London: Town and Country Planning Association.

Criekingen, M. van and Decroly, J. (2003) 'Revisiting the diversity of gentrification: neighbourhood renewal processes in Brussels and Montreal', *Urban Studies*, 40(12): 2451–68.

Crook, T., Currie, J., Jackson, A., Monk, S., Rowley, S., Smith, K. and Whitehead, C. (2002) 'Planning gain and affordable housing', York: Joseph Rowntree Foundation.

Danielzyk, R. (1992) 'Gibt es im Ruhrgebiet eine "postfordistische Regionalpolitik"?', *Geographische Zeitschrift*, 2: 84–105.

Davies, A. and Ford, S. (1999) 'Art futures', *Art Monthly*, 223: 9–11.

Davies, T. (1999) 'Amsterdam: comments on a city of culture', Amsterdam: Afdeling Kunst en Cultuur.

Deutsche, R. (1996) *Evictions: Art and Spatial Politics*, Massachusetts: MIT Press.

Development Trusts Association (2007) *What is a development trust?* Online. Available HTTP: http://www.dta.org.uk/aboutourmembers/whatisadevelopmenttrust.htm (accessed 10 October 2007).

Donald, B. (2002) 'Spinning Toronto's golden age: the making of a "city that worked"', *Environment and Planning A*, 34(12): 2127–54.

Donnison, D. and Middleton, A. (eds) (1987) *Regenerating the Inner City: Glasgow's Experience*, London: Routledge and Kegan Paul.

Draaisma, J. and Hoogstraten, van P. (1983) 'The squatter movement in Amsterdam', *International Journal of Urban and Regional Research*, 7(3): 405–16.

Duivenvoorden, E. (2000) 'Een voet tussen de deur, geschiedenis van de kraakbeweging 1964–1999. Uitgeverij De Arbeiderspers', Amsterdam: Antwerpen.

Ekonomist (2007) 'TOKI'nin Istanbul planı: Istanbul'da 1 milyon Ev Yıkılacak' (The MHA's Istanbul Plan: one million houses will be demolished in Istanbul), *Kasım* 45: 19–26.

Elias, N. and Scotson, J.L. (1994) *The Established and the Outsiders: A Sociological Enquiry into Community Problems*, London: Sage Publications.

Ercan, F. and Oguz, S. (2007) 'Rethinking anti-neoliberal strategies through the perspective of value theory: Insights from the Turkish case', *Science and Society*, 71(2): 173–202.

Esser, J. and Hirsch J. (1989) 'The crises of fordism and the dimension of a "post-Fordist" regional and urban structure', *International Journal of Urban and Regional Research*, 13(3): 417–37.

Evans, G. (2001) *Cultural Planning: An Urban Renaissance?*, London: Routledge.

—— (2005) 'Measure for measure: Evaluating the evidence of culture's contribution to regeneration', *Urban Studies*, 42(5/6): 959–83.

Fainstein, S. (1993) *The City Builders*, Oxford UK: Blackwell

Fernandes, A. and Filgueiras Gomes, M.A.A. de (1995) 'Pelourinho: turismo, identidade e consumo cultural', in M.A. de Filgueiras Gomes (ed.) *Pelo Pelò: historia, cultura e cidade*, Salvador: Editora da UFBA-Facultade de Arquitetura-Maestrado em Arquitetura e Urbanismo.

Ferrara, E. (2008) 'Dieci milioni in più per il parcheggio', *Repubblica cronaca di Firenze*, 9 January: 91.

Filey, M. (1996) *Remember Sunnyside: The Rise and Fall of a Magical Era*, Toronto: Dundurn Press.

Firenze 2010 (2001) 'Progettare Firenze: Materiali per il piano strategico dell'area metropolitana', Firenze: Comune network.

Florida, R. (2002) *The Rise of the Creative Class: And How It's Transforming Work, Leisure, Community and Everyday Life*, New York: Basic Books.

Foment de Ciutat Vella, S.A. (2006a) 'Memòria de Foment de Ciutat Vella SA 2005', Barcelona: Foment de Ciutat Vella, S.A.

—— (2006b) 'Programa per a l'adquisició de locals comercials tancats al Districte de Ciutat Vella', Barcelona: Foment de Ciutat Vella, S.A.

Friedli, B., Pöhner, R. and Widmer, T. (1997) 'Die nebensächlichste Hauptstadt der Welt', *Facts*, 28 August. Online. Available HTTP: http://www.angelfire.com/ms/zschokke/bern. html (accessed 30 January 2008).

Friedmann, J. (2000) 'The good city: In defence of utopian thinking', *International Journal of Urban and Regional Research*, 24(2): 460–72.

—— (2007) 'The wealth of cities: Towards an assets-based development of newly urbanizing regions', *Development and Change*, 38(6): 987–98.

Gans, H. (1968) *People and Plans: Essays on Problems and Solutions*, London: Basic Books.

Ganser, K., Siebel, W. and Sieverts, T. (1993) 'Die planungsstrategie der IBA Emscher Park. eine Annäherung', *Raumplanung*, 61: 7–31.

García-Peralta, B. (2005) 'Gestión gubernamental de la producción habitacional en México, 1930–2000', unpublished thesis, Universidad Nacional Autónoma de México.

—— (2006) 'Housing for the working class on the periphery of Mexico City: A new version of gated communities', *Social Justice*, 33(3): 129–41.

Garcia-Ramon, M.D. and Albet, A. (2000) 'Pre-Olympic and post-Olympic Barcelona, a "model" for urban regeneration today?', *Environment and Planning A*, 32(8): 1331–34.

Garza Villarreal, G. (ed.) (2000) *La ciudad de México en el fin del segundo milenio*, Mexico City: El Colegio de México.

—— (2006) 'Macroeconomía de la ciudad de México', paper presented at Urban Age Conference, Mexico City, 23–25 February. Online. Available HTTP: http://www.urban-age.net (accessed 30 November 2007).

Gevisser, M. (2004) 'From the ruins: The Constitution Hill project', *Public Culture*, 16(3): 507–19.

Ginsburg, N. (1999) 'Putting the social into urban regeneration policy', in *Local Economy*, 14(1): 55–71.

Glass, R. (1963) *Introduction to London: Aspects of Change*, London: Centre for Urban Studies.

Gobierno del Distrito Federal (GDF) (2000) 'Bando Informativo 2. Se restringe el crecimiento de unidades habitacionales y desarrollos comerciales en las Delegaciones Álvaro Obregón, Coyoacán, Cuajimalpa de Morelos, Iztapalapa, Magdalena Contreras, Milpa Alta, Tláhuac, Tlalpan y Xochimilco', Mexico City: GDF.

Goldman, L. and The Red Room (2005) *Hoxton Story: An interactive archive, a booklet and a limited series of intimate walkabout performances*. Online. Available HTTP: http://www.theredroom.org.uk/hoxton.htm (accessed 27 August 2007).

Gomez, P. (2008) 'Il tram e la cupola', *L'espresso*, 14 February. Online. Available HTTP: http//:espresso.repubblica.it/dettaglio/Il-tram-e-la-cupola/1988767//0 (accessed 20 February 2008).

Gordon, I. (2004) 'The resurgent city: what, where, how and for whom?', *Planning Theory and Practice*, 5(3): 371–79.

Gordon, I. and Buck, N. (2005) 'Cities in the New Conventional Wisdom', in N. Buck, I. Gordon, A. Harding and I. Turok (eds) *Changing Cities: Rethinking Urban Competitiveness, Cohesion and Governance*, Basingstoke: Palgrave Macmillan.

Gotham, K.F. (2005) 'Tourism gentrification: the case of New Orleans' Vieux Carre (French Quarter)', *Urban Studies*, 42(7): 1099–121.

Government Office for the South East, East Midlands, East of England (2005) 'Milton Keynes and South Midlands sub-regional strategy', London: The Stationary Office.

Graham, N. and Roemer, D. (2007) 'Last call at the Gladstone Hotel', documentary, Toronto: Produced by Derreck Roemer and Neil Graham.

Greenall, R. (2000) *A History of Northamptonshire*, Chichester: Phillimore and Co.

Grup de patrimoni industrial del Fòrum Ribera Besòs (2006a) *Proposta de criteris d'intervenció*, Barcelona: Fòrum Ribera Besòs. Online. Available HTTP: http://www.salvemcanricart.org (accessed 13 July 2007).

—— (ed) (2006b) 'Can Ricart: Patrimoni, innovació i ciutadania', Barcelona: Fòrum Ribera Besòs.

—— (2006c) 'Can Ricart, recinte o fragments? Tractament del patrimoni, inserció urbana i reutilització del complex fabril', Barcelona: Forum Ribera Besòs. Online. Available HTTP: http://www.salverncanricart.org (accessed 13 July 2007).

Grup d'Etnologia dels Espais Públics de l'Institut Català d'Antropologia (2006) *Victimes del 22@... les empreses de Can Ricart*. Online. Available HTTP: http://barcelona.indymedia.org/usermedia/image/12/large/victimes22@_can_ricart__s.jpg (accessed 10 July 2007).

GYODER (2007) *7. Gayrimenkul Zirvesi Sonuç Bildirgesi* (The Final Declaration of Real Estate Summit 7 on April 25–27). Online. Available HTTP: http://www.gyoder.org.tr/zirve7/index.htm (accessed on 29 May 2008).

Hackworth, J. (2002) 'Postrecession gentrification in New York city', *Urban Affairs Review*, 37(6): 815–43.

Hackworth, J. and Smith, N. (2001) 'The changing state of gentrification', *Tijdschrift voor Economische en Sociale Geografie*, 92(4): 464–77.

Hall, C. M. (2005) 'Seducing global capital – reimaging space and interaction in Melbourne and Sydney', in C. Cartier and A.A. Lew (eds) *Seductions of Place: Geographical Perspectives on Globalization and Touristed Landscapes*, London: Routledge.

Hämer, H.-W. (ed) (1984) 'Idee, Prozess, Ergebnis. Die Reparatur und Rekonstruktion der Stadt; Internationale Bauausstellung Berlin 1987', [Senator für Bau- und Wohnungswesen], Berlin: Frölich & Kaufmann.

—— (1990) 'Behutsame Stadterneuerung', in Senatsverwaltung für Bau- und Wohnungswesen (ed.) *Stadterneuerung Berlin*, Berlin: Senatsverwaltung.

Hämer, H.-W. (2007) *The Cautious City* [interview by H. Karssenberg], Erasmus PC. Online. Available HTTP: http://www.erasmuspc.com/index.php?id=18108&type=article (accessed 23 December 2007).

Hamnett, C. (2003) *Unequal City: London in the Global Arena*, London: Routledge.

Hansdampf (1998) 'Reithalle Bern: Autonomie und Kultur im Zentrum', Zürich: Rotpunktverlag.

Harvey, D. (1985) *The Urbanization of Capital: Studies in the History and Theory of Capitalist Urbanisation*, Oxford: Basil Blackwell.

—— (1989a) *The Urban Experience*, Oxford: Blackwell

—— (1989b) 'From managerialism to entrepreneurialism: The transformation in urban governance in late capitalism', *Geografiska Annaler B*, 71(1): 3–17.

—— (1989c) *The Condition of Postmodernity*, Oxford: Blackwell.

—— (1997) *La crisi della modernità*, Milano: Il Saggiatore.

—— (2000) *Spaces of Hope*, Edinburgh: Edinburgh University Press.

—— (2001) 'The art of rent: Globalization and the commodification of culture', in D. Harvey (ed.) *Spaces of Capital: Towards a Critical Geography*, Edinburgh: University of Edinburgh Press.

Herman, M. and Leuthold, H. (2002) *The Consequences of Gentrification and Marginalisation on Political Behaviour*, Zürich: Institute of Geographie, University of Zürich. Online. Available HTTP: www.sotomo.geo.unizh.ch/research/ (accessed 29 May 2008).

—— (2004) 'Bern hat Basel links überholt', *Der Bund*, 7 December: 23.

Hitz, H.R., Keil, R., Lehrer, U., Ronneberger, K., Schmid, C. and Wolff, R. (eds) (1995) *Capitales Fatales: Urbanisierung und Politik in den Finanzmetropolen Frankfurt und Zürich*, Zürich: Rotpunkt.

Hoffmann-Axthelm, D. (1987) *Baufluchten: Beiträge zur Rekronstruktion der Geschichte Berlin-Kreuzbergs*, Berlin: Transit.

Hoffmann-Nowotny, H.J. (1973) *Soziologie des Fremdarbeiterproblems: eine theoretische und empirische Analyse am Beispiel der Schweiz*, Stuttgart: Ferdinand Enke.

Homuth, K. (1984) *Statik Potemkinscher Dörfer: Anmerkungen zum Verhältnis von "behutsamer Stadterneuerung" und gesellschaftlicher Macht in Berlin-Kreuzberg*, Berlin: Ökotopia.

Hoskins, G.C. and Tallon, A.R. (2004) 'Promoting the urban idyll: Policies for city centre living', in C. Johnstone and M. Whitehead (eds) *New Horizons in British Urban Policy: Perspectives on New Labour's Urban Renaissance*, Aldershot: Ashgate.

Hume, C. (2005) 'Bohemian rhapsody: The Gladstone Hotel's lifesaving renovation is the latest in the ongoing hipsterization of Queen West', *Toronto Star*: A3.

IBA (1999*) Internationale Bauausstellung Emscher Park, IB '99 Finale*, Gelsenkirchen: IBA Emscher Park.

Imbroscio, D. (2008) '"[U]nited and actuated by some common impulse of passion": Challenging the dispersal consensus in American housing policy research', *Journal of Urban Affairs*, 30(2): 111–30.

Imrie, R. and Raco, M. (2003) 'Community and the changing nature of urban policy', in R. Imrie and M. Raco (eds) *Urban Renaissance?: New Labour, Community and Urban Policy*, Bristol: The Policy Press.

Initiativkreis Emscherregion (1994) *Zum Stand der Dinge: Strukturwandel im Ruhrgebiet. Dialoge zur regionaneln Entwicklung*, Dortmund: Initiativkreis Emscherregion.

Initiativkreis Ruhrgebiet (2007) *Contractfuture*. Online. Available HTTP: http://www. contractfuture.de/project/project.jsp (accessed 28 October 2007).

Inner-City Community Forum (2003) *Johannesburg: Interfund Development Update. Johannesburg's Better Buildings Programme: A Response*. Online. Available HTTP: http:// www.interfund.org.za/pdffiles/vol5_one/InnerCity.pdf (accessed 3 November 2004).

Instituto Nacional de Estadística Geografía e Informática (INEGI) (1970) 'Censo de población y vivienda', Mexico City: INEGI.

—— (1980) 'Censo de población y vivienda', Mexico City: INEGI.

—— (1990) 'Censo de población y vivienda', Mexico City: INEGI.

—— (1995) 'Censo de población y vivienda', Mexico City: INEGI.

—— (2000) 'Censo de población y vivienda', Mexico City: INEGI.

—— (2005) 'II conteo de población y vivienda 2005', Mexico City: INEGI.

Islam, T. (2005) 'Outside the core: gentrification in Istanbul', in R. Atkinson and G. Bridge (eds) *Gentrification in a Global Context: The New Urban Colonialism*, London: Routledge.

Italian National Institute of Statistics (2001) 'National demographic census', Rome: Italian National Institute of Statistics.

Jacobs, J. (1961) *The Death and Life of Great American Cities*, New York: Random House.

Johnson, B. and Homan, S. (2003) 'Vanishing acts – an inquiry into the state of live popular music opportunities in New South Wales', Sydney: Australia Council and the NSW Ministry for the Arts.

Jones, M. and Ward, K. (2004) 'Neo-liberalism, crisis and the city: the political economy of New Labour's urban policy', in C. Johnstone and M. Whitehead (eds) *New Horizons in British Urban Policy: Perspectives on New Labour's Urban Renaissance*, Aldershot: Ashgate.

Kabisch, S. (2001) 'Wenn das kleid der stadt nicht mehrpasst – strategien im Umgang mit dem Wohnungs leerstand in ostdeutschen Städtenaus stadtplanerischer und stadtsoziologischer perspective', UFZ-Diskussi onspapiere Number 3, Leipzig: UFZ.

Kamsma, T. (1998) 'Amsterdam terug in de europese toerisme-topvijf, de herontdekking van de jeugd als topattractie', in I. van Eerd (ed.) *Pluriform Amsterdam: Essays*, Amsterdam: Vossiuspers AUP.

Kempen van, R. and Priemus, H. (2002) 'Revolution in social housing in the Netherlands: Possible effects of new housing policies', *Urban Studies*, 39(2): 237–53.

Kerrigan, M. (2006) 'Higher education in the East Midlands: A widening participation perspective. Aim higher', Loughborough: Loughborough University.

Keyder, C (2005) 'Globalization and social exclusion in Istanbul', *International Journal of Urban and Regional Research*, 29(1): 124–34.

Kilper, H. and Wood, G. (1995) 'Restructuring policies: The Emscher Park International Building Exhibition', in P. Cooke (ed.) *The Rise of the Rustbelt*, London: UCL Press.

Kipfer, S. and Keil, R. (2002) 'Toronto.Inc?: Planning the competitive city in the New Toronto', *Antipode*, 34(2): 227–64.

Knapp, W. (1998) 'The Rhine–Ruhr area in transformation: Towards a European metropolitan region?', *European Planning Studies*, 6(4): 379–93.

Krätke, S. (2004) 'City of talents?: Berlin's regional economy, socio-spatial fabric and "worst practice" urban governance', *International Journal of Urban and Regional Research*, 28(3): 511–29.

Krätke, S. and Borst, R. (2000) *Berlin: Metropole Zwischen Boom und Krise*, Opladen: Leske & Budrich.

Kunzmann, K. (1999) 'White work elephants in the Ruhr district's park of the future', *TOPOS – the European Landscape Magazine*, 26 March: 79–86.

Kurtuluş, H. (2007) 'Global flows and new segregation in Istanbul: How cloudy class identities have crystallized in urban space', in R. Sandhu and J. Sandhu (eds) *Globalizing Cities: Ineqality and Segregation in Developing Countries*, New Delhi: Rawal Books.

Laboratorio sulle scelte urbanistiche nel I Municipio (2003) *Qualità e vivibilità nel centro storico: Idee e proposte per una politica urbana*, Rome: self-published.

Landry, C. (1996) *The Art of Regeneration: Urban Renewal through Cultural Activity*, Stroud: Comedia.

—— (2000) *The Creative City: A Toolkit for Urban Innovators*, Stroud: Comedia.

Landry, C. and Bianchini, F. (1995) *The Creative City*, London: Demos.

Lange, B. (2007) 'Die Räume der Kreativeszenen: culturepreneurs und ihre orte in Berlin', Bielefeld: transcript.

Larner, W. (2003) 'Neoliberalism?', *Environment and Planning D*, 21(5): 509–12.

Lees, L. (1994) 'Rethinking gentrification: Beyond the positions of economics or culture', *Progress in Human Geography*, 18(2): 137–50.

—— (2000) 'A reappraisal of gentrification: Towards a "geography of gentrification"', *Progress in Human Geography*, 24(3): 389–408.

—— (2003a) 'Super-gentrification: The case of Brooklyn Heights, New York City', *Urban Studies*, 40(12): 2487–509.

Lees, L., Slater, T. and Wyly, E. (2008) *Gentrification*, London: Routledge.

Lefebvre, H. (2003) *The Urban Revolution*, translated by Robert Bononno with a Foreword by Neil Smith, Minneapolis: University of Minnesota Press.

Leggett, T. (2003) 'Rainbow tenement: crime and policing in inner Johannesburg', *Monograph*, 78: 8–33.

Lehrer, U. (2006) 'Re-placing Canadian cities: the challenge of landscapes of "desire" and "despair"', in T. Bunting and P. Filion (eds) *The Canadian City in Transition*, 3rd edn, Oxford: Oxford University Press.

Lehrer, U. and Laidley, J. (2006) 'Old mega-projects newly repackaged? Waterfront redevelopment in Toronto', International Sociological Association, Conference Proceedings, Durban, South Africa, 22–29 July.

Lehrer, U. and Winkler, A. (2006) 'Public or private? The Pope Squat and housing struggles in Toronto', *Social Justice: A Journal of Crime, Conflict, and World Order*, 33(3): 142–58.

Lehrer, U. with Alkasawat, L., Berard, E., De Gregorio, E., Di Carlo, M., Di Clemente, L., Fancello, E., Fisher, K., Garcia, R., Harris, J., Hodge, M., Huq, M., Robinson, S., Stevens, K., Sloly, D. and Vaiya, S. (2006) 'The West Queen West Triangle: Report', Faculty of Environmental Studies, York University, unpublished, 116 pp.

Levitt, J. and Adams, D. (2005) 'Reinvention by design: an examination of the role of design in the ongoing evolution of Queen Street West', PowerPoint Presentation. Online. Available HTTP: http://active18.org (accessed 1 September 2006).

Ley, D. (1994) 'Gentrification and the politics of the new middle class', *Environment and Planning D*, 12(1): 53–74.

—— (1996) *The New Middle Class and the Remaking of the Central City*, Oxford: Oxford University Press.

—— (2003) 'Artists, aestheticisation and the field of gentrification', *Urban Studies*, 40(12): 2527–44.

Ley, D. and Mills, C. (1993) 'Can there be a postmodernism of resistance in the urban landscape?', in P. Knox (ed.) *The Restless Urban Landscape*, New Jersey: Prentice Hall.

Lindner, R. (2007) 'The creativity hype and the culture of cities', paper presented to the Liquid Cities symposium, University of Western Sydney, 4–5 October.

Lipietz, A. (1986) 'New tendencies in the international division of labour: Regimes of accumulation and modes of regulation', in A. Scott and M. Storper (eds) *Production, Work, Territory: The Geographical Anatomy of Industrial Capitalism*, Boston: Allen & Unwin.

Live Music Taskforce (2003a) 'Options paper', Melbourne: Department of Sustainability and Environment, 2 October.

—— (2003b) 'Report and Recommendations', Melbourne: Government of Victoria, December.

Logan, W. (1985) *The Gentrification of Inner Melbourne: A Political Geography of Inner City Housing*, Australia: University of Queensland Press.

Lucas, C and Collins, S. (2007) '"Very Melbourne" kind of shopping idea in the works', *The Age*, 21 November: 5.

Lupi, T. (2007) 'Van probleemwijk naar krachtstad', *Agora*, 23(3): 4–8.

Lüthi, C. and Meier, B. (1998) *Bern – eine Stadt bricht auf: Schauplätze und Geschichten der Berner Stadtentwicklung zwischen 1798 und 1998*, Bern, Stuttgart, Wien: Haupt.

Lütke-Daldrup, E. and Doehler-Bhezadi, M. (2004) *PlusMinus. Leipzig 2030 – Stadt in Transformation. Transforming the City*, Wuppertal: Müller + Busmann KG.

McNeill, D. (2003) 'Mapping the European urban left: The Barcelona experience', *Antipode*, 5(1): 74–94.

MacLeod, G. and Ward, K. (2002) 'Spaces of utopia and dystopia: Landscaping the contemporary city', *Geografiska Annaler B*, 84(3–4): 153–70.

Marcuse, P. (1985) 'Gentrification, abandonment and displacement: Connections, causes and policy responses in New York City', *Journal of Urban and Contemporary Law*, 28: 195 240.

Marshall, T. (ed.) (2004) *Transforming Barcelona*, London: Routledge.

Martin, A. (ed.) (2007) 'Diagnòstic Socioeconòmic Ambiental del Casc Antic: Document I', Barcelona: ARDA Gestió i Estudis Ambientals.

Mas, M. and Verger, T. (2004) 'El forat de la vergonya al casc antic de Barcelona', in Unió Temporal d'Escribes (eds) *Barcelona Marca Registrada: un Model per a Desarmar*, Barcelona: Virus Editorial.

Media Spree (2007) 'Wasser in der stadt', Berlin: Media Spree.

Memoli, M. (2005) *La città immaginata: Spazi sociali, luoghi, rappresentazioni a Salvador de Bahia*, Milano: Franco Angeli.

Meyer, C. (2006) 'Media spree: The epicentre of Berlin's music industry', in U. Reinhard (ed.) *taktvoll 2.0*, Berlin: whois.

Miles, M. (2005) 'Interruptions: Testing the rhetoric of culturally led urban development', *Urban Studies*, 42(5/6): 889–911.

Miles, S. (2005) '"Our Tyne": Iconic regeneration and the revitalisation of identity in Newcastle Gateshead', *Urban Studies*, 42(5/6): 913–26.

Miles, S. and Paddison, R. (2005) 'Introduction: The rise and rise of culture-led urban regeneration', *Urban Studies*, 42(5/6): 833–39.

Milroy, S. (2005) 'A fresh bloom on the Queen West scene', *The Globe and Mail*, 28 January.

Ministério das Cidades (2007) *Assessoria de Comunicação*. Online. Available HTTP: http://www.cidades.gov.br/index.php?option=content&task=view&id=2927 (accessed 10 January 2008).

Ministerium für Stadtentwicklung, Wohnen und Verkehr (ed.) (1988) 'Emscher Park: werkstatt für die zukunft alter industriegebiete, memorandum zu inhalt und organisation', Düsseldorf: Minister für Stadtentwicklung, Wohnen und Verkehr des Landes NRW.

Ministry of Economic Affairs and Energy of the State of North Rhine–Westphalia (2006) 'Transformation through cultural (industries) in the Ruhr: culture (industries) through transformation. Essen for the Ruhr: a contribution to Essen's bid to be the 2010 European capital of culture', Essen: Kulturhauptstadt Europas 2010.

Montgomery, J. (1995) 'Urban vitality and the culture of cities', *Planning Practice and Research*, 10(2): 101–09.

Newman, K. and Wyly, E. (2006) 'The right to stay put, revisited: Gentrification and resistance to displacement in New York city', *Urban Studies*, 43(1): 23–57.

Nigg, H. (ed.) (2001) *Wir wollen alles, und zwar subito!: die Achtziger Jugendunruhen in der Schweiz und ihre Folgen*, Zürich: Limmat.

North Northants Joint Planning Unit (2007) 'North Northamptonshire core spatial strategy', Corby: Joint Planning Unit.

O + S Amsterdam (2007), 'Bureau voor onderzoek en statistiek', Amsterdam: Onderzoek en Statisktiek.

Odendaal, N. (2005) '[D]urban space as the site of collective action: Towards a conceptual framework for understanding the digital city in Africa', paper presented at the Planning and Natural Resource Management Conference, eThekwini Municipality, Durban, 24–28 May.

Office of the Deputy Prime Minister (2003) *The Sustainable Communities Plan*, London: HMSO.

Oliva, A. (2003) *El districte d'activitats 22@bcn*, Barcelona: Aula Barcelona.

Özdemir, N. (1996) 'Transformation of squatter settlements by the public sector: The case of Dikmen Valley, Ankara', in R. Camstra and J. Smith (eds) *Housing: Levels of Perspective*, Amsterdam: AME.

Paba, G. (ed.) (2001) *Insurgent City: Racconti e Geografie di un Altra Firenze*, Livorno: Media Print.

Parkinson, M., Hutchins, M., Simmie, J., Clark, G. and Verdonk, H. (2004) *Competitive European Cities: Where do the Core Cities Stand?*, London: ODPM.

Peck, J. (2005) 'Struggling with the creative class', *International Journal of Urban and Regional Research*, 29(4): 740–70.

Peck, J. and Tickell, A. (2002) 'Neoliberalizing space', *Antipode*, 34(3): 380–404.

Perrons, D. and Skyers, S. (2003) 'Empowerment through participation? Conceptual explorations and a case-study', *International Journal of Urban and Regional Research*, 27(2): 265–85.

PINA (2007) 'Yasadıgımız 50 yılın Ardından, Cogunlugun Cıkarlarını Savunan ve Bizi Insan Yerine Koyan Planlama Surecleri Istiyoruz (After our experience of past 50 years, we are claiming the planning processes advocating the interest of majority and treating us as the people)', *Planlama*, 38: 103–11.

Planungs- und Wirtschaftsdirektion (1970) 'Modellstudie Lorraine: möglichkeiten einer quartiererneuerung', Bern: Planungs- und Wirtschaftsdirektion.

Plug Cultura (2007) *Reforma do Centro Histórico entra em nova etapa*. Online. Available HTTP: http://www.cultura.ba.gov.br/noticias/plugcultura/reforma-do-centro-historico-entra-em-nova-etapa (accessed 6 March 2008).

PMB Amsterdam (2000) 'Geen cultuur zonder subcultuur, plan van aanpak broedplaats, vastgesteld door de gemeenteraad 21 juni 2000', Amsterdam: Gemeente Amsterdam.

Porter, L. and Barber, A. (2006) 'Closing time: The meaning of place and state-led gentrification in Birmingham's Eastside', *City*, 10(2): 215–34.

—— (2007) 'Planning the cultural quarter in Birmingham's Eastside', *European Planning Studies*, 15(10): 1327–48.

Porter, L. and Hunt, D. (2005) 'Birmingham's Eastside story: Making steps towards sustainability?', *Local Environment*, 10(5): 525–42.

Pratolini, V. (1943) *Il quartiere*, Firenze: Vallecchi.

Pruijt, H. (2003) 'Is the institutionalization of urban movements inevitable? A comparison of the opportunities for sustained squatting in New York City and Amsterdam', *International Journal of Urban and Regional Research*, 27(1): 133–57.

Pyner, M. (2007) Personal communication from the Executive Director of the Shoreditch Trust, London, 17 September.

Rada, U. (1997) *Hauptstadt der Verdrängung*, Berlin & Göttingen: Schwarze Risse – Rote Straße & Libertäre Assoziation.

Rampini, F. (2007) 'L'assedio alla pechino antica', *La Domenica di Repubblica*, 19 December: 28.

Real Estate Institute of Victoria (2004) 'Property update', Melbourne: Real Estate Institute of Victoria, June.

Riemens, P. and Lovink, G. (2002) 'Local networks: Digital city Amsterdam', in S. Sassen (ed.) *Global Networks, Linked Cities*, London: Routledge.

Roberts, P. and Sykes, H. (eds) (2000) *Urban Regeneration: A Handbook*, London: Sage.

Robson, G. and Butler, T. (2001) 'Coming to terms with London: Middle-class communities in a global city', *International Journal of Urban and Regional Research*, 25(1): 70–86.

Rose, D. (1984) 'Rethinking gentrification: Beyond the uneven development of Marxist urban theory', *Environment and Planning D*, 2(1): 47–74.

Rose, D. (1996) 'Economic restructuring and the diversification of gentrification in the 1980s: A view from a marginal metropolis', in J. Caulfield and L. Peake (eds) *City Lives and City Forms: Critical Research and Canadian Urbanism*, Toronto: University of Toronto Press.

Sandercock, L. (1998) *Towards Cosmopolis: Planning for Multicultural Cities*, Chichester: John Wilcy and Sons.

—— (2003) *Cosmopolis II: Mongrel Cities in the 21st Century*, London: Continuum.

Santos, M. (1995) 'Salvador: centro e centralidade na cidade contemporanea', in M. A. de Filgueiras Gomes (ed.) *Pelo Pelò: historia, cultura e cidade*, Salvador: Editora da UFBA-Facultade de Arquitetura-Maestrado em Arquitetura e Urbanismo.

Sartogo, V. (2007) 'Homo movens: il dramma dei trasporti', in AA.VV. (ed.) *Modello Roma: L'ambigua modernità*, Rome: Odradek Edizioni.

Sartorio, S.F. (2005) 'Strategical spatial planning', *disP*, 162(3): 26–40.

Sassen, S. (1994) *Cities in a World Economy*, Thousand Oaks: Pine Forge Press.

Scarpaci, J. (2000) 'Reshaping Habana Vieja: Revitalisation, historic preservation, and restructuring in the socialist city', *Urban Geography*, 21(8): 724–44.

Scharenberg, A. (2000) *Berlin: Global City oder Konkursmasse?*, Berlin: Karl Dietz.

Scharenberg, A. and Bader, I. (eds) (2005) *Der Sound der Stadt: Musikindustrie und Subkultur in Berlin*, Münster: Westfälisches Dampfboot.

Scott, A.J. (1997) 'The cultural economy of cities', *International Journal of Urban and Regional Research*, 21(2): 323–39.

Scott, A. and Storper, M. (eds) (1986) *Production, Work, Territory: The Geographical Anatomy of Industrial Capitalism*, Boston: Allen & Unwin.

Scott, J.C. (1998) *Seeing Like a State: How Certain Schemes to Improve the Human Condition have Failed*, New Haven, CT: Yale University Press.

Seager, A. (2007) 'House builders win battle against green technologies', *The Guardian*, 20 August: 22.

Searle, G. (2002) 'Uncertain legacy: Sydney's Olympic stadiums', *European Planning Studies*, 1(10): 845–60.

Searle, G. and Bounds M. (1999) 'State powers, state land and competition for global entertainment: The case of Sydney', *International Journal of Urban and Regional Research*, 23 (1):165–72.

Secretaría de Desarrollo Urbano y Vivienda (2001) *Programa General de Desarrollo Urbano del Distrito Federal, Mexico City: SEDUVI*. Online. Available HTTP: http://www.seduvi.df.gob.mx/programas/pgdu.html (accessed 28 January 2008).

Selvatici, F. (2008a) 'Ghiberti, tutti a processo', *Repubblica cronaca di Firenze*, 8 February: 9.

—— (2008b) 'La procura indaga sui tre parcheggi costruiti con il project financing', *Repubblica cronaca di Firenze*, 19 February: 4.

Senatsverwaltung für Stadtentwicklung (2007) *Stadtentwicklungsplanung – Entwicklung der Berliner Wasserlagen. Kreuzberger und Friedrichshainer Spreeufer*. Online. Available HTTP: http://www.stadtentwicklung.berlin.de/planen/stadtentwicklungsplanung/ de/wasserlagen/raeume/spreeufer.shtml (accessed 23 December 2007).

Şengül, T. (1999) 'The Turkish earthquake: An end to the "neoliberal state"?', *Third World Planning Review*, 21(4): iii–viii.

—— (2003) 'On the trajectory of urbanisation in Turkey: An attempt at periodisation', *International Development Planning Review*, 25(2): 153–68.

Seo, J.-K. (2002) 'Re-urbanisation in regenerated areas of Manchester and Glasgow', *Cities*, 19(2): 13–121.

Shaw, K. (2000) 'Whose image? Global restructuring and community politics in the inner-city', unpublished Master's thesis, RMIT University, Melbourne.

—— (2002) 'Culture, economics and evolution in gentrification', *Just Policy*, 28(December): 42–50.

—— (2005a) 'Local limits to gentrification: implications for a new urban policy', in R. Atkinson and G. Bridge (eds) *Gentrification in a Global Context: The New Urban Colonialism*, London: Routledge.

—— (2005b) 'The place of alternative culture and the politics of its protection in Berlin, Amsterdam and Melbourne', *Planning Theory and Practice*, 6(2): 149–69.

—— (2006) 'The trouble with the creative class', *Planning News*, 32(1): 4–5.

Shaw, S.J. and MacLeod, N.E. (2000) 'Creativity and conflict: Cultural tourism in London's city fringe', *Tourism, Culture and Communication*, 2(3): 165–75.

Shoreditch Trust (2007) *Delivery Plan 2007/08*, London: Shoreditch Trust. Online. Available HTTP: http://www.shoreditchtrust.org.uk (accessed 27 August 2007).

Silimela, Y. (2003) *Johannesburg: Interfund Development Update. Urban Renewal and City Reconstruction: Government Position*. Online. Available HTTP: http://www.interfund. org.za/vol5no12004.html (accessed 3 November 2004).

Slater, T. (2004a) 'Municipally managed gentrification in South Parkdale, Toronto', *Canadian Geographer*, 48(3): 303–25.

—— (2004b) 'North American gentrification? Revanchist and emancipatory perspectives explored', *Environment and Planning A*, 36(7): 1191–213.

—— (2006) 'The eviction of critical perspectives from gentrification research', *International Journal of Urban and Regional Research*, 30(4): 737–57.

Smith, N. (1987) 'Of yuppies and housing: Gentrification, social restructuring and the urban dream', *Environment and Planning D*, 5(2): 151–72.

—— (1996) *The New Urban Frontier: Gentrification and the Revanchist City*, London: Routledge.

—— (2002a) 'New globalism, new urbanism: Gentrification as global urban strategy', *Antipode*, 34(3): 427–50.

—— (2002b) 'New globalism, new urbanism: Gentrification as global urban strategy', in N. Brenner and N. Theodore (eds) *Spaces of Neoliberalism: Urban Restructuring in North America and Western Europe*, Oxford: Backwell.

Soja, E. (2000) 'The stimulus of a little confusion, a contemporary comparison of Amsterdam and Los Angeles', in L. Deben (ed.) *Understanding Amsterdam: Essays on Economic Vitality, City Life and Urban Form*, 2nd edn, Amsterdam: Het Spinhuis.

Sokoloff, B. (1999) *Barcelone, ou Comment Refaire une Ville*, Montreal: Les Presses Universitaires de Montréal.

Springer, B. (2006) 'Art as a medium of urban upgrading: "The Heeresbäckerei" in Berlin-Kreuzberg contrasted to Zim in Rotterdam', *Tijdschrift voor Economische en Sociale Geografie*, 97(5): 610–16.

Stadtforum (2006) 'Aktuelle fragen und probleme der Leipziger stadtentwicklung', Leipzig: Stadtforum Leipzig.

Stadtplanungsamt, B.S. (1982) 'Inventar Lorraine', Bern: Stadtplanungsamt.

Statistikdienste der Stadt Bern (2006) 'Sozialraumanalysen 1990/2000', Bern: Statistikdienste.

Stahl, G. (2004) '"It's like Canada reduced": Setting the scene in Montreal"', in A. Bennett and K. Kahn-Harris (eds) *After Subculture: Critical Studies in Contemporary Youth Culture*, UK: Palgrave MacMillan.

Stienen, A. (2006) 'Verborgene Einschluss- und Ausgrenzungsdynamik im Stadtteil', in A. Stienen (ed.) *Integrationsmaschine Stadt?: Interkulturelle Beziehungsdynamiken am Beispiel von Bern*, Bern, Stuttgart, Wien: Haupt.

—— (2007) 'Sozialräumliche Stadtentwicklung: Eine Interpretation der Sozialraumanalysen am Beispiel ausgewählter Quartiere', Bern: Statistikdienste.

Subirats, J. and Rius, J. (2004) 'Del Xino al Raval: cultura i transformación social a la Barcelona central', Barcelona: Centre de Cultura Contemporània de Barcelona.

Sudjic, D. (2008) 'Keeping cities alive', *The Age*, 13 May: 11.

Swyngedouw, E. (2007) 'Impossible "sustainability" and the postpolitical condition', in R. Krueger and D. Gibbs (eds) *The Sustainable Development Paradox: Urban Political Economy in the United States and Europe*, London: The Guilford Press.

Tamayo, S. (ed.) (2007) 'Los desafíos del Bando 2: Evaluación multidimensional de las políticas habitacionales en el Distrito Federal 2000–2006', Mexico City: SEDUVI.

Taylor, R. (2005) 'Fashion victims', *The Guardian*, 7 September. Online. Available HTTP: http://society.guardian.co.uk/communities/story/0,1563742,00.html (accessed 27 August 2007).

The Wildlife Trust for Birmingham and the Black Country and University of Birmingham (2005) *Eastside Biodiversity Strategy*, Birmingham. Online. Available HTTP: http://www.sustainable-eastside.net/eastside_activities.html (accessed 13 March 2008).

Tillim, G. (2005) *Cape Town: Michael Stevenson Gallery*. Online. Available HTTP: http://www.michaelstevenson.com/contemporary/exhibitions/jhb/jhb1.htm (accessed 7 December 2005).

Touraine, A. (1985) 'An introduction to the study of social movements', *Social Research*, 52(4): 749–88.

Town and Country Planning Association (2007) 'Best practice in urban extensions and new settlements', London: Town and Country Planning Association.

Transform! Italia (2005) *La riva sinistra del Tevere: Mappe e conflitti nel territorio metropolitano di Roma*, Rome: Transform! Ed.

Treanor, P. (2002) *Brownfield Gentrification*. Online. Available HTTP: http://web.inter.nl.net/users/Paul.Treanor/gasfab (accessed May 2003).

Tripodi, L. (2004) 'The abrogated city', in INURA (ed.) *The Contested Metropolis: Six Cities at the Beginning of 21st Century*, Basel: Birkhauser.

Trullen, J. (2000) 'El projecte Barcelona ciutat del coneixement, des de l'economia: Barcelona Metròpolis Mediterrània (1)', Barcelona: Ajuntament de Barcelona.

Tuckett, I. (1988) 'Coin Street: There is another way . . .', *Community Development Journal*, 23(4): 249–57.

Turok, I. (2005) 'Cities, competition and competitiveness: Identifying new connections', in N. Buck, I. Gordon, A. Harding and I. Turok (eds) *Changing Cities: Rethinking Urban Competitiveness, Cohesion and Governance*, Basingstoke: Palgrave Macmillan.

Urban Dictionary (2007) *Shoreditch twat*. Online. Available HTTP: http://www.urbandictionary.com/define.php?term=Shoreditch+twat&defid=1121974 (accessed 27 August 2007).

Uitermark, J. (2004) 'Framing urban injustices: The case of the Amsterdam squatter movement', *Space and Policy*, 8(2): 227–44.

Verschueren, J. (2003) 'Plaats voor de homo ludens!', unpublished thesis, Department of Modern History, University of Gent, Gent.

Vicario, L. and Monje, P.M.M. (2005) 'Another "Guggenheim effect"?: The generation of a potentially gentrifiable neighbourhood in Bilbao', *Urban Studies*, 40(12): 2383–400.

Victorian Government (2003) 'Taskforce to find solutions to live music battle: Media Release', Melbourne: Arts Victoria, 6 June.

—— (2004a) 'Victoria planning provisions', Melbourne: Department of Sustainability and Environment.

—— (2004b) 'Live music: The way forward. Supporting a sustainable live music scene in Victoria', Melbourne: Department of Sustainability and Environment.

VicUrban (2008) *Melbourne Docklands: Statistics & Projections*. Online. Available HTTP: http://www.docklands.com.au (accessed 10 January 2008).

Villaça, F. (2001) *Espaço intra-urbano no Brasil*, Sao Paulo: Studio Nobel-FAPESP-Lincoln Institute.

Voss, A. (2007) 'The Emscher Park International Building Exhibition: A trendsetting model?', in P. Oswalt (ed.) *Shrinking Cities*, Berlin: Project Office Philipp Oswalt.

Wakelin, M. (2004) 'Beware investment apartments', *The Age* Business section, 11 July: 3.

van Weesep, J. (1994) 'Gentrification as a research frontier', *Progress in Human Geography*, 18(1): 74–83.

Wilson, D., Wouters, J. and Grammenos, D. (2004) 'Successful protect-community discourse: Spatiality and politics in Chicago's Pilsen neighbourhood', *Environment and Planning A*, 36(7): 1173–90.

Winkler, T. (2006) 'Kwere Kwere journeys into strangeness: Reimagining inner city regeneration in Hillbrow, Johannesburg', unpublished thesis, University of British Columbia, Vancouver, Canada.

Wirtschaftsmagazin Ruhr (2007) *Wirtschaftsmagazin Ruhr*, 3(5).

WirtschaftsWoche and IW Consult (2007) 'Grosser Staedtetest', *WirtschaftsWoche*, 37: 26–38.

Wolff, R. (1998) 'A star is born', in INURA (ed.) *Possible Urban Worlds: Urban Strategies at the End of the 20th Century*, Basel: Birkhäuser.

Wyly, E. and Hammel, D. (2005) 'Mapping neo-liberal American urbanism', in R. Atkinson and G. Bridge (eds) *Gentrification in a Global Context: The New Urban Colonialism*, London: Routledge.

Yeoh, B. (2004) 'Cosmopolitanism and its exclusions in Singapore', *Urban Studies*, 41(2): 2431–45.

Ziegler, Maya (2002) 'Marginalisierung peripherer arbeiterquartiere: Schleichender soziostruktureller wandel in städtischen aussenquartieren und seine auswirkungen auf die weltanschauung aufgezeigt am beispiel Zürich, Bern, Winterthur', unpublished Masters dissertation, Geographisches Institute, Universität Zürich. Online. Available HTTP: http://www.sotomo.geo.unizh.ch/research/masters/msc_ziegler.pdf (accessed 27 May 2008).

Žižek, S. (1998) *Pleidooi voor intolerantie*, Amsterdam: Boom Essays.

Zukin, S. (1982) *Loft Living: Culture and Capital in Urban Change*, Baltimore, MD: Johns Hopkins University Press.

—— (1995) *The Cultures of Cities*, Cambridge, MA: Blackwell.

Zukin, S. and Kosta, E. (2004) 'Bourdieu off-Broadway: Managing distinction on a shopping block in the East Village', *City and Community*, 3(2): 101–14.

Index